THE BELIEVING WARRIOR

Why Believers Should Serve in the Military

———————

Nahum Meléndez, PhD, MDiv, BCC-APC

DEDICATION

I would like to dedicate this book to my wife, Veronica, and my daughter, Dalía, whom have been my motivation into writing this piece. Also, to my supportive family who have always supported and encouraged me.

The Believing Warrior: Why Believers Should Serve in the Military.
Copyright © 2025 by Nahum I Meléndez

All rights reserved. No part of this book may be reproduced or transmitted in any form or by any means without written permission from the author.

Published by Nahum Meléndez.
Printed by Lulu.com
ISBN 978-1-7365704-0-1

Table of Contents

PREFACE

INTRODUCTION

PART I—The Joining Debate
    CHAPTER 1: Should An Immigrant Join the Armed Forces?
    CHAPTER 2: Pros & Cons of Becoming a servicemember
    CHAPTER 3: The Opposition Says…
    CHAPTER 4: God May Have a Plan for You
    CHAPTER 5: To Kill or Not to Kill?

PART II—The Elite Forces in the Bible
    CHAPTER 6: Are Armed Forces Biblical?
    CHAPTER 7: Biblical Combat in the Bible
    CHAPTER 8: Military Training vs Spiritual Training
    CHAPTER 9: What is the Job of a Warrior?

PART III—The Preparation Task
    CHAPTER 10: Boot Camp is Not for Cowards
    CHAPTER 11: "Get Up and Walk"
    CHAPTER 12: The Conversion Experience

PART IV—Going to War
    CHAPTER 13: The Morse Code
    CHAPTER 14: Combat Strategies
    CHAPTER 15: The Merciless Enemy Attacks

PREFACE

IT STARTS WITH A QUESTION. One that echoes through midnight phone calls, nervous living rooms, and quiet, prayerful tears: *Can a believer serve in the military and still honor God?* I've heard it spoken through trembling lips and bold ones. I've heard it from teenage dreamers and weary parents. I've heard it from those already in uniform and those standing at the crossroads, unsure of which path to take.

This question rarely comes from a place of rebellion. More often, it rises from a place of restless integrity. A young man or woman, deeply convicted, hungry for meaning, and wrestling with what it means to be both a soldier and a servant of Christ. They ask:

- *Should I join the military as a believer?*
- *What if I am given a weapon and have to kill someone?*
- *Can military take away my freedom of religion?*
- *Will God approve learning martial arts and combat techniques?*
- *Would I betray my family's faith by choosing to serve?*

I've looked into the eyes of those asking, and I've seen in them a younger version of myself—uncertain, searching, torn between a desire to serve and the fear of abandoning sacred convictions. These are not trivial questions. They deserve more than surface-level answers. They deserve the voice of someone who has stood at that same threshold.

Short Biography

I have served now for almost 13 years, both, as a military enlisted in the United States Marine Corps and now as an officer in the United States Navy. Each of these experiences shaped my worldview and who I am today as a minister, a scholar, a son, a husband and father, but also as a citizen of this great nation.

I have served this country for nearly fifteen years—first as an enlisted Marine and now as a Navy officer and chaplain. These experiences have shaped more than my career. They've shaped my worldview, my theology, my fatherhood, my manhood, and my calling as a minister. The military did not strip me of my faith—it refined it. It didn't erase my convictions—it tested them, burned them, and rebuilt them stronger.

My journey was not one of pristine clarity. I wasn't a bright-eyed seminary graduate the day I raised my right hand. I was a college dropout, a young man with a wrecked first car, a heart full of anger, and an addiction ready to swallow me whole. My parents were divorcing. My faith was fraying. I was desperate for direction, aching for identity, and chasing anything that felt like control. I didn't join the Marines to be heroic. I joined to survive.

But what I didn't realize then—and what I understand now—is that God had written Himself into my orders. That Parris Island wasn't a detour from ministry, but the beginning of it. That in a world filled with spiritual confusion and moral compromise, the battlefield would become my classroom, my sanctuary, and my pulpit.

This book exists because no one gave me a faithful, grounded, biblically sound, real-world answer to the questions I once carried. And I promised God—if He ever gave me clarity, I would give it to others. That's what you're holding now: not just a book, but a conversation between me and every warrior wondering if they've been called.

Let me briefly tell you where this conviction has taken me. I earned a Ph.D. in Human Development, writing my dissertation on moral injury—specifically among trauma nurses who, like combat veterans, carry wounds deeper than what flesh can show. I completed Clinical Pastoral Education

(CPE) under the Association for Clinical Pastoral Education (ACPE), a board that recognizes me today as a Board-Certified Chaplain.[1] I've been entrusted with guiding people through death, grief, addiction, suicide, faith crises, and impossible decisions—both in and out of uniform.

My education spans theology, psychology, ministry, and leadership. From Puerto Rico to Michigan to military installations across the country, I've poured myself into the pursuit of truth, compassion, and purpose. I've served as a parish pastor, earned a Master's in Divinity, and am nearing the completion of an MBA. But none of this would have happened if I hadn't made one scared, half-surrendered decision to walk into a recruiting office and say, "I need something different."

God took that moment—not my strength, not my discipline—and began to mold something holy from it.

How This Book Came to Be

This book was born over years of study, deployment, failure, restoration, and reflection. It is both practical and theological, deeply personal and unapologetically biblical. My goal is not to argue that all Christians should join the military. But I will argue—firmly, faithfully, and with conviction—that if God calls you there, you must not refuse Him simply because it makes others uncomfortable.

We begin not with policy or politics—but with prophecy.

Revelation 12 tells of a war that did not start on Earth, but in Heaven. There, Michael and his angels fought against the dragon. Evil was not born on the battlefield—it was born in rebellion against God's order. And from that war came exile. From that exile came deception. And from that deception, the dragon now wages war on the "remnant of her seed"—those who keep God's commandments and hold to the testimony of Jesus.

You cannot read Revelation 12 and remain spiritually neutral. You are *either in the war or unaware of it.*

---

[1] Association of Clinical Pastoral Education, "Accreditation," accessed November 30, 2020, https://acpe.edu/programs/accreditation.

7 And there was war in heaven: Michael and his angels fought against the dragon; and the dragon fought and his angels,

8 And prevailed not; neither was their place found any more in heaven.

9 And the great dragon was cast out, that old serpent, called the Devil, and Satan, which deceiveth the whole world: he was cast into the earth, and his angels were cast out with him.

10 And I heard a loud voice saying in heaven, Now is come salvation, and strength, and the kingdom of our God, and the power of his Christ: for the accuser of our brethren is cast down, which accused them before our God day and night.

11 And they overcame him by the blood of the Lamb, and by the word of their testimony; and they loved not their lives unto the death.

12 Therefore rejoice, ye heavens, and ye that dwell in them. Woe to the inhabiters of the earth and of the sea! For the devil is come down unto you, having great wrath, because he knoweth that he hath but a short time.

13 And when the dragon saw that he was cast unto the earth, he persecuted the woman which brought forth the man child.

14 And to the woman were given two wings of a great eagle, that she might fly into the wilderness, into her place, where she is nourished for a time, and times, and half a time, from the face of the serpent.

15 And the serpent cast out of his mouth water as a flood after the woman, that he might cause her to be carried away of the flood.

16 And the earth helped the woman, and the earth opened her mouth, and swallowed up the flood which the dragon cast out of his mouth.

17 And the dragon was wroth with the woman, and went to make war with the remnant of her seed, which keep the commandments of God, and have the testimony of Jesus Christ.

- Kings James Version

Here are reasons why YOU SHOULD read on:

#1 Reason. If you are a young person wondering whether or not the military is for you, this is the book for you. We will explore the advantages and disadvantages of the military from someone who is not afraid of sharing the biological, psychological, social and spiritual impact the military can have on you.

#2 Reason. If you are a parent whose child has approached you with such curiosity and you are concerned about whether the military will test your child's faith and perhaps stray him or her away from their faith, then read on.

#3 Reason. If you are a pastor, elder, church officiant, neighbor, friend, or significant other and someone approached you with such questions and you have no idea how to respond, this book will help you consider alternative answers. Perhaps, this is the same child who is afraid asking his or her own parents for fear they will say, "no" without first understanding this militant construct.

Here are reasons why YOU SHOULD NOT read on:

#1 Reason. If you are completely opposed to military service due a number of reasons: lack of exegetical knowledge, someone with only enlisted service has influenced you, you have preconceived biases inherited from your own family's concept of evil and war; or just plain opposed to learn about the topic.

#2 Reason. If you do not anticipate anyone will ever ask you about the ethical and moral implications of joining the military as a God-fearing person or if you are just plain old and do not think you can influence future generations.

While my hope is to speak to those living in North America, the principles, lessons, and personal experiences shared in this book transcend these borders. These are real struggles, real truths and real lives that can speak to a wider youth community with or without faith affiliation facing one of the most important decisions; to join the military.

INTRODUCTION

THERE HAS ALWAYS BEEN tension—holy tension—between faith and warfare, between the sword and the sanctuary, between the battlefield and the altar. And for generations, God-fearing people have wrestled with one enduring question: *Can I serve both God and my country without betraying either?*

This question has filled countless pages of books, sermons, articles, and theological debates. But if I'm honest, many of those voices sound hollow to those of us who have worn the uniform. They speak in theory, not experience. They argue from afar, not from the trenches. And while their intentions may be sincere, their conclusions often leave deep wounds—especially when they discredit the very believers who have laid their lives on the line in places most pastors will never walk.

I've spent hours, perhaps too many, reading what I call *comfortable critiques*—those who write with moral clarity but little military insight. Among these was a book titled *I Pledge Allegiance*, a well-meant attempt to discourage Christian service in the armed forces. But as I turned its pages, a sadness settled in. Not because the author didn't care, but because his message painted every Christian soldier, sailor, airman, and Marine as spiritually compromised. His conclusions, drawn from a short stint in uniform during the 1980s, seem to eclipse the complexity of calling—ignoring the very real moral courage it takes to serve both Christ and country in a time of war.

One of the authors, a civilian seminary student and PhD candidate in systematic theology, writes from an academic vantage point. The other, a pastor who completed a single seven-year enlistment, speaks from his limited time in uniform. To their credit, both aim to protect young believers from the spiritual compromises they themselves encountered or feared. I respect their transparency and their concern for the next generation. But I believe they've missed something vital—something you only discover when you've served long enough to see God at work in the trenches. God doesn't retreat from the world—He invades it.

He doesn't avoid darkness—He sends His light straight into it. He doesn't shelter His people from evil—He raises up warriors to confront it. The idea that the military is too secular, too broken, or too violent to be touched by believers is not only a narrow reading of Scripture—it's a denial of redemption itself.

What if God needs warriors in every sector of society—yes, even in the military? What if the Marine standing post, the sailor swabbing the deck, or the chaplain in combat boots is *the very vessel God uses to reach the lost, comfort the broken, and restrain the tide of evil?*

Isn't that what we see in Scripture?

From Joseph in Pharaoh's court to Daniel in Babylon's palace, from Esther in the Persian empire to the centurion at the foot of the cross—God has always placed His people inside powerful, dangerous, and even morally complicated systems not to blend in, but to stand out. He never told them to wait for sinners to wander into synagogues. He said, *"Go into all the world."* Russell Burrill put it this way: *"The early disciples were not to wait for people to come to them—they were to go to people with the message God had given them. The marching order for God's church was 'Go.'"*[1]

And so I went.

I didn't begin this book with the intent of building an apologetic for military service. I'm not here to argue, debate, or convince those who've already made up their minds. Rather, I write because I've lived on both sides of the uniform—in the foxhole and the pulpit. I've stood at attention and knelt in prayer. I've fired a weapon and lifted trembling

---

[1] Russell Burrill, *Reaping the Harvest: A Step-by-step Guide to Public Evangelism* (Fallbrook, Calif.: Hart Books, 2007), 17.

hands to bless the broken. And what I've discovered is this: God does not abandon those who serve. He often calls them first.

This book is not just theology. It's testimony. It's the story of how military service saved my life—not just physically, but spiritually, emotionally, and morally. It's how God used one of the most feared militant institutions in the world, the United States Marine Corps, to carve purpose into a lost young man's heart.

My Testimony

In November of 2003, I enlisted in the United States Marine Corps—easily one of the most elite and ferocious fighting forces on the planet. At just over two hundred thousand active-duty Marines, this branch has been rightly called "America's 911." We were the first in, often the last out. We were trained for the world's hardest missions, summoned to restore order where chaos reigned. And I was one of them.

But that's not where my story begins. The real story began long before Parris Island. I was 21 years old, directionless, heartbroken, drifting in a fog of mistakes. I had dropped out of college. My first car had been totaled. My soul was spiraling toward addiction. My family was collapsing under the weight of a painful divorce. I had lost my sense of who I was—and more frighteningly—who I might become. I was a son without vision. A man without a mission.

And then, one unexpected day, the call to serve showed up like a flare in the night sky. Not from a church altar, but from a recruiter's desk. It wasn't a spiritual moment—at least not outwardly. But I believe now that God was in it from the start. I needed discipline, brotherhood, courage, conviction, and a purpose bigger than myself. And I found all of that…in the uniform of a Marine.

Four years later, in November of 2007, I left active duty. But I didn't leave the mission. I just changed battlefields. I transitioned from warfighter to pastor—from a battalion to a congregation. I traded my rifle for a Bible, my body armor for a shepherd's heart. And through that transformation, I realized that the war was never just external—it was always spiritual.

What happened in those four years that changed me? What did the Marine Corps awaken in me that years of church attendance hadn't? Why did I leave as one man and return as another? These are the questions I'll answer in the pages ahead—not as a polished theologian, but as a wounded warrior redeemed by grace.

What You Will Discover

This book is raw and redemptive. It will not shy away from hard truths or difficult memories. It will show you the spiritual cost of service, the mental toll of combat training, the moral questions that never come with easy answers. But it will also reveal how God never abandoned me in the process. He was in the barracks. In the field. In the dark nights of doubt. He was in the discipline. In the loss. In the quiet prayers of a young man trying to hold onto his faith while learning how to fight.

You will walk with me through moments of spiritual crisis and personal breakthrough. You will see how God used the crucible of the Corps to refine my heart. You may even see yourself in these pages—your own battles, your own brokenness, your own questions. You might cry. You might rejoice. You might find, like I did, that God sometimes sends us into the military not because He's forgotten us—but because He's preparing us.

So read this not as an argument, but as an offering. If my journey can give hope to someone struggling with the same questions, then this book has done its job.

PART I

THE JOINNING DEBATE

CHAPTER 1

Should An Immigrant Join the Armed Forces?

I WAS ABOUT TWENTY ONE-YEAR-OLD when I joined the Marine Corps. I had only been living in the state of Maryland for about five years, which meant that I had only been learning the English language for that amount of time and was not as proficient as I would have wanted to. I still wrestled with the culture shock and getting used to the American ways.

Even though I had finished three years of high school and I could understand parts of a conversation with someone, yet I was still not in the condition to join a militia force. Military was not an option for me. My raw English disqualified me to be a military man, let alone a Marine. I could barely understand the actors speaking on the movies let alone understand a command from a drill instructor, or worse, a call for help in the battlefield. I was terrified with the thought that a life would depend on me and find myself unable to save it because of my language barrier.

Feelings of Inferiority as an Immigrant

As a Hispanic I was doomed to do what everybody else was doing: construction, painting, plumbing, gardening, maintenance, dish washing, waiter and other low paying jobs that nobody wanted. Do not misunderstand me; I am not saying these jobs make someone less of a person, in the contrary, these jobs may mold your character in a way no other would do. They can teach you humility and, in practical way, they can save you a lot of money in the long run if you learn them correctly.

This was my case working with my father in the painting business. I learned to sand wood, scrape old paint, mix different colors, set up the dust cloth around the painting site, use cocking to fill the cracks, pull out nails, work with plaster, write out job proposals, etc. I learned a great deal of manual work, which I will be eternally thankful to my father for teaching me these skills.

But to an immigrant like me, filled with dreams and higher expectations, this was not something I saw myself doing for the rest of my life. With a country filled with endless opportunities, having immigrant parents, where one is working in the painting-carpentry field and the other in the bus school system, my siblings and I did not have much of a future in the academic world since we could not afford to go to an university.

What really held me back was the fear that I would not make it as an immigrant. I now realize that many immigrants like me suffer from this sense of inferiority, special among the first and second-generation immigrants. They feel like they do not belong to their mother country because they are living in the United States now and they also feel they do not belong to this country because they came from another.

The Inclusion of Immigrants in the Bible

This was the case of many immigrants in the Bible. Immigration goes back since Adam and Eve were forced to migrate out of the Garden of Eden and settling in the east (Gen. 3:22-24). Later, the people at the Babel migrated to other places (Gen. 11:1). Centuries later, Abraham and all his family migrated to the land of Canaan (Gen. 11:31-12:9). Since then, God's people have always been migrating from one place to the other. In the New Testament, God tells his new Christian church to migrate and proclaim the gospel (Mat. 28:16-20).

According to the Department of Homeland Security, there are approximately 11.5-20 million illegal immigrants in United States.[1] This constitutes a 4 to 6.5 percent of the total community of three hundred and eight million Americans. One of the biggest ethical issues for the church in the immigration debate is whether or not the church should be involved among the politic discussions or continue to take a philanthropic lawlessness stand towards the issue, which for some zealots it is just a fancy way to translate treason.[2]

Some believe that the New Testament fails to treat this subject explicitly; however, all throughout the Bible there is a thematic posture on how to treat immigrants and how to integrate them in our community. We should explore this ethical issue by asking: who is considered an immigrant? How were immigrants integrated in the Biblical time? To answer the first question, we will analyze the Greek term ἔθνη "gentile" as well as other Greek terms to describe who was an immigrant. To answer the latter question, we will analyze the treatment of Jesus with some immigrants and what He said about immigrant integration. Finally, we will provide a rather short analysis to what it means to γίνομαι "become" a πολιτείας "citizen."

From the very beginning of human history, movement has been both consequence and calling. Adam and Eve migrated because of sin; Abraham migrated because of faith. Israel migrated because of bondage; the church migrated because of mission. Every migration carried a purpose larger than geography—it was about identity, obedience, and the fulfillment of divine purpose.

---

[1] Michael Hoefer, Nancy Rytina, and Bryan Baker, "Estimates of the Unauthorized Immigrant Population Residing in the United States: January 2011," http://www.dhs.gov/xlibrary/assets/statistics/publications/ois_ill_pe_2011.pdf, June 23, 2013, accessed June 23, 2013.

[2] Michael L. Budde, The Borders of Baptism: Identities, Allegiances, and the Church (Eugene, OR: Wipf & Stock Pub, 2011), 85.

In the same way, today's immigrants—whether crossing borders or cultures—stand at the crossroads of purpose and belonging. The Bible reminds us that those who are displaced are not disqualified; rather, they are often the very instruments God uses to advance His plan. Abraham left his homeland to become the father of nations; Ruth left Moab to preserve a royal lineage; even Jesus "migrated" from heaven to earth for the salvation of humanity.

This divine pattern of movement reflects the heart of The Believing Warrior. Just as God calls immigrants to serve His mission through faith, He also calls believers—citizens of heaven yet residents of earthly nations—to serve through courage, discipline, and sacrifice. The immigrant who seeks belonging in a new homeland mirrors the believer who seeks to live faithfully in the world while belonging to another Kingdom.

In both cases, there is a sacred tension between citizenship and service. The believer-soldier, like the faithful immigrant, learns that true allegiance is not divided but deepened through commitment—to God first, and to the mission He has entrusted us with on earth. The same virtues that define the faithful migrant—courage, endurance, loyalty, and hope—are the very virtues that shape a believing warrior.

Immigration and Integration in the New Testament

The New Testament provides at least seven variations in translations of an immigrant: (1) ξένων which is translated by the KJV as "strangers" in Rom. 16:23. (2) παροίκους and (3) παρεπιδήμους which is translated by NIV as "foreigners" and "exiles" in 1 Pet. 2:11. (4) ἀλλοτρίων which is translated by the NRSV as "others" or "alien" in Mat. 17:25 and Eph 2:12. (5) ἐπιδημοῦντες which is translated by the NASB as "proselyte" in Act. 2:10. (6) παροικίας which is translated by the WYC as "pilgrim" in 1 Pet. 1:17. (7) ἐθνῶν which is translated by most Bible translations as "gentile" in Mat. 4:15 except the KJV 1900 which also translates it as "nation" in Mat. 28:19.

In each one of these instances the author is referring to a group of people or a person outside of the Jew community or not belonging to this specific group. For the sake of this book, we will only analyze four key passages that deals with the treatment of an immigrant.

In the New Testament, words like *xenos, paroikos,* and *ethnos* reveal that believers are spiritual foreigners—citizens of heaven living in earthly lands. The early church understood this tension between belonging and mission. Likewise, the soldier of faith serves in a world not his own, loyal both to God and to the nation he protects. Just as Gentiles were once outsiders but became *sympolitai*—fellow citizens through Christ—believers learn that true citizenship requires sacrifice and discipline. The integration of immigrants mirrors the believer's integration into God's kingdom: both demand courage, obedience, and service. The believing warrior embodies this dual identity—rooted in heavenly allegiance yet committed to earthly duty—showing that serving in the military can be an act of faith as much as of patriotism.

Mathew 15:21-28—

A lot of Matthew's focus is discipleship. According to Campbell, Matthew, as disciple himself, focuses on portraying Jesus as the King, a call to the lifestyle a disciple must follow and a charge on how anybody can become a disciple.[3]

First of all, Matthew starts telling the reader about the genealogy of Jesus in which several outsiders or strangers made a significant contribution within God's redemptive plan. In chapter 1 he tells us about Tamar, a Canaanite woman, who belonged to the ancient Palestines;[4] Rahab who was a gentile prostitute mother of Boaz citizen of the Jericho city;[5] Ruth, a Moabite,[6] who according to Nehemiah were considered a foreigner to Israelites and banned from the assembly of God's people (Deut. 23:3-4).

---

[3] Iain D. Campbell, *Matthew's Gospel (opening Up)* (Leominster: DayOne Publications, 2008), overview section.

[4] Merriam-Webster, Inc. Merriam-Webster's Collegiate Dictionary. Eleventh ed. Springfield, MA: Merriam-Webster, Inc., 2003. 983.

[5] Ibid., 870.

[6] Ibid., 895

[7] Ibid., 1061.

Then, chapter 2 continues with the mention of some men from the east, who were considered wise and of noble status,[7] to be part of the worship experience of the new Almighty Baby. It is clear that for Matthew the issue of immigration and integration is of major concern. We can also see this in the subsequent chapters which shows Jesus teaching about the love we should have towards our enemies which are described as ἐθνικός "gentiles" (5:47). In this instance, Jesus commands his disciples to not only love your own ἀδελφός or fellow Jew, but also the ἐθνικός by integrating them in their society.

Finally, we get to chapter 15 and Jesus encounters a Canaanite woman who had a daughter who was possessed by a demon. Interesting enough Matthew is really trying to convey the idea that although Jesus was a Jew, for God there is no favoritism and everybody is welcome to enjoy His power. This idea is shown when Jesus' family came to see Him claiming a special privilege with Jesus, but His response is that those who do God's will are to be considered His family and not Mary and His brothers only (12:46-50).

Later, Matthew carefully adds the story of the false teaching of the Scribes and the Pharisees followed by the explanation of what it means to have a pure heart and abruptly he introduces the story of this gentile, the Canaanite woman in v.21, who ὅριον ἐκείνων ἐξέρχομαι "came out of those borders." According to Carroll, an immigrant is a person who moves across borders;[8] therefore, this unknown woman whom Matthew fails to give any more description other than that she was an immigrant, but highlights the interaction between Jesus and this immigrant. He hears the request this woman whose faith was μέγας "greater" than any Jew. In v.23 we see the disciples, the church, rejecting this immigrant in contrast with the accepting of Jesus in v.25-28.

Matthew's account of the Canaanite woman reveals that faith and perseverance often emerge from the margins. Her boldness to cross borders—social, cultural, and spiritual—embodies the courage of those willing to fight for what is right. Just as Jesus honored her μέγας (great)

---

[8] M. Daniel Carroll R, *Christians at the Border: Immigration, the Church, and the Bible* (Grand Rapids, Mich.: Baker Academic, 2008), 65.

faith, the believing warrior is called to display conviction under pressure and humility in service. In both discipleship and military life, loyalty and endurance define true strength. The woman's story teaches that God's favor is not bound by ethnicity or status, but by obedience and persistence. Likewise, the believer who serves in uniform demonstrates that faith knows no borders—discipleship and duty can coexist when both are rooted in a heart willing to cross boundaries for the sake of righteousness and compassion.

John 1:1-3; 4:1-30—

In the Book of John, Jesus is portrayed as the λόγος who migrated to this earth for the benefit of humanity. His intent is to relate Jesus' work as an immigrant among humans and how, as an immigrant, suffered rejection, need and even death by the society in which He lived.
John tells us that the λόγος ἐγένετο "became" or "came into existence" as a Jew citizen. This word ἐγένετο is later used by Paul in Eph. 2:13 to describe how a gentile becomes a citizen of the Kingdom of Jesus. It seems that the same meaning of divine becoming or ἐγένετο has to take place in order to be part of Jesus' kingdom which is of heavenly nature according to Jesus (Jn. 18:36).

It is through a miraculous transformation that one becomes part of this Kingdom. In addition, John infers in order to be a citizen of this Kingdom one must ἐγένετο "come into existence," "cross the borders" of our own limitations and enter in this realm of Spiritual residency; although this may not always be pleasant and may cause you harshness.

While Matthew is appealing to the acceptance of immigrants, John is appealing to the need to become an immigrant for the sake of others. In chapter 4:1-30 an interaction between Jesus and a Samaritan woman seems for John necessary to be included right after John the Baptist finishes explaining about the need for the migration of Jesus from heaven to this world. While John emphasizes the need for immigration and how important and beneficial it is to welcome immigrants Jesus emphasizes the importance of integration.

In v.4, the Greek word δεῖ which its literal translation is "it was necessary"—for Jesus it was necessary to migrate to Samaria (same Greek word is used in 3:7, 14, 30). John makes a special emphasis in Jesus ministry. Carroll talks about this instance and says that this road was the road required to journey north, but another level of necessity is at work [in Jesus' ministry]. This occasion was an important part of what Jesus needed to do for his mission. This meeting is not an unplanned, coincidental happenstance, but rather part of the predetermined plan of Jesus.[9]

Again, integration is seen by the interaction between Jesus and this gentile woman. Another aspect of this Jew Jesus is that He takes the initiative to mingle with this gentile despite the prejudices raised by the society, hence the reason why the disciples become surprised when they saw Jesus mingling with a gentile in v.27.

John's Gospel presents Jesus as the Logos—the divine Word who willingly "migrated" from heaven to earth to redeem humanity. His incarnation was not a coincidence but a mission. The use of the word ἐγένετο ("became") captures a divine transformation—God crossing the greatest border between the infinite and the finite. This act of leaving comfort for the sake of others mirrors the heart of every believing warrior who serves beyond self-interest. Jesus' journey through Samaria was described as δεῖ—"it was necessary." His crossing of social and cultural barriers demonstrates intentional obedience to purpose, even in the face of misunderstanding.

Likewise, the military believer learns that service is often uncomfortable, demanding sacrifice, endurance, and compassion. When Jesus engaged the Samaritan woman, He modeled a warrior's courage with a servant's heart. Her transformation reminds us that authentic mission happens when one dares to bridge division. The believing warrior, like Christ, must cross boundaries—between faith and duty, heaven and earth—to bring healing where others see hostility. In this divine migration, service becomes sacred; battle becomes redemptive; and every act of courage becomes an echo of Christ's self-giving mission to reconcile all

---
[9] Ibid., 118-119.

people to God.

Luke 17:11-18—

Now Luke brings in the personal attitude an immigrant must take towards the society that welcomes him. In other words, not only should the Jews be responsible for treating well the foreigners, but now this same immigrant must also be thankful for such kindness. Samaritans were gentiles. Carroll gives us a very good description about them. He says that they practiced a form of Judaism, by they had a separate holy mountain (Mount Gerizim), their own priesthood, and special beliefs and rituals. They were not accepted as equals by other Jews, and the antipathy between them ran deep.[10]

In this passage, again we see Jesus accepting the immigrant, the Samaritan who was considered a ἀλλογενής "foreigner," by healing him from leprosy. Interesting enough, this immigrant ὑπέστρεψεν "returned" which is also mentioned by Jesus in v.18 emphasizing that an immigrant must also be thankful of those who open their doors to provide for them. Coincidentally, ὑπέστρεψεν has also a connotation of worship as described in Luke 2:20; 17:15; 24:52, pointing out that thankfulness and praise to God may always be intertwined.

Luke's account of the ten lepers, especially the Samaritan who returned, shifts the focus from societal duty to personal gratitude and transformation. Among the ten who were healed, only one—a foreigner—came back to give thanks. This act of returning, expressed in the Greek ὑπέστρεψεν ("returned"), carries both physical and spiritual meaning. It was not merely a turn of the body but a turning of the heart toward the One who gave him new life. Gratitude became the immigrant's act of worship, a visible declaration that mercy must be met with devotion. In this, Luke emphasizes that true healing is not complete until it produces thankfulness. The Samaritan's faith transcended his social label; he was no longer a marginalized ἀλλογενής (foreigner) but a participant in divine grace. His response models the believer's proper attitude toward both

---
[10] Ibid., 117.

God and the communities that welcome them—an attitude rooted in humility, acknowledgment, and service.

For the believing warrior, gratitude is the foundation of discipline and duty. Just as the Samaritan returned to honor his Healer, the Christian soldier is called to return daily in recognition of God's sustaining power and protection. Military life often demands sacrifice, endurance, and resilience—qualities shaped by a grateful heart. Gratitude transforms service from mere obligation into sacred stewardship. The healed Samaritan reminds us that faith is not passive acceptance but active acknowledgment—a returning to the source of all strength. Likewise, the soldier who serves with gratitude honors not only their nation but their Creator, finding worship in every act of duty. In the rhythm of obedience, thanksgiving becomes the believer's uniform, distinguishing those who fight for righteousness from those who fight for recognition. The believing warrior, like the thankful Samaritan, turns every moment of service into a living testimony of grace, loyalty, and divine gratitude.

Ephesians 2:11-22—

Although there is no variants in the Ephesians 2:11-13, we find that ποτὲ "once" could also be understood as "used to be" as translated by the Good News Translation in Eph. 5:8 which denotes a time that has passed before. The Greek New Testament hints that people ἀπαλλοτριόω "alienated from" were to receive the ἐπαγγελία "promise" given to Abraham.[11] For Paul, an immigrant or gentile who accepts the sacrifice of Jesus Christ becomes a citizen of the heavenly Kingdom.

11 Διὸ μνημονεύετε ὅτι ποτὲ ὑμεῖς τὰ ἔθνη ἐν σαρκί, οἱ λεγόμενοι ἀκροβυστία ὑπὸ τῆς λεγομένης περιτομῆς ἐν σαρκὶ χειροποιήτου,
12 ὅτι ἦτε τῷ καιρῷ ἐκείνῳ χωρὶς Χριστοῦ, ἀπηλλοτριωμένοι τῆς πολιτείας τοῦ Ἰσραὴλ καὶ ξένοι τῶν διαθηκῶν τῆς ἐπαγγελίας, ἐλπίδα μὴ ἔχοντες καὶ ἄθεοι ἐν τῷ κόσμῳ.

---

[11] Kurt Aland et al., eds., *The Greek New Testament, 4th Revised Edition*, 4 Revised ed. (Stuttgart, Germany: American Bible Society, 2000), 658.

CHAPTER 1: SHOULD AN IMMIGRANT JOINT THE ARMED FORCES

13 νυνὶ δὲ ἐν Χριστῷ Ἰησοῦ ὑμεῖς οἵ ποτε ὄντες μακρὰν ἐγενήθητε ἐγγὺς ἐν τῷ αἵματι τοῦ Χριστοῦ.

My Translation

In v. 11-13, I have chosen to translate *ethnos* as immigrants to the fact that it is usually translated as "gentiles" in the New Testament. "Gentiles" are referred to as "outsiders" throughout the gospel. I have also translated *apallotrioo* as "noncitizen" since it means to be non-participant of a community; in this case community is the Jews. In addition, *xenos* is yet another term to refer to immigrants since the literal translation is "strangers" or "foreigners." Finally, *ginomai* can also mean "come into existence" which I believe best fit our text based on the context:

11 Therefore, keep remembering that you who used to be *immigrants* in the flesh—those once called "the uncircumcised" by those who are circumcised by human hands
12 used to be without the Messiah, *noncitizens* of the commonwealth of Israel, and *foreigners* to the covenants of the promise. You lived without hope and without God in this world.
13 But now, through the blood of Christ Jesus, you who used to be far away have *come into existence* as those brought near—reborn into citizenship through His sacrifice.

In the context of The Believing Warrior, this passage parallels the process of enlistment: leaving behind a civilian identity to adopt a disciplined, purposeful life of service. The believer, once an outsider, now bears the mark of belonging to a divine command. Citizenship in God's kingdom is not inherited—it is earned through grace, sealed in blood, and lived out through duty. The warrior's transformation mirrors the immigrant's—both find new identity through commitment, sacrifice, and allegiance to a greater cause.

Word Study

"Therefore" This is the third major truth of Paul's doctrinal section (chapters 1–3). The first was God's eternal choice based on His gracious character, the second was *the hopelessness of fallen humanity, saved by God's gracious acts through Christ* which must be received and lived out by faith. Now the third, *God's will has always been the salvation of all humans* (cf. Gen. 3:15), both Jew and Gentile (cf. 2:11–3:13).

These Gentiles are commanded to continue to remember (*"remember"* is a present, active, imperative) their previous alienation from God, vv. 11–12. Paul obviously understood Jesus' mission on earth as described in Eph. 2:11-22 as for he himself became an immigrant and understood what it meant to be a spiritual foreigner. He commands to "remember" that Christians were once ἔθνος "gentiles" and called ἀκροβυστία "uncircumcised" by those who were circumcised by human hand. This type of remembering in to be done again and again, constantly how through the Lord we are no longer ἀπαλλοτριόω "non-participants" of Israel and ξένος "strangers" or "immigrants," but we are now συμπολίτης "fellow citizens" of the οἰκεῖος τοῦ θεός "members of the household of God."

This remembering is to give immigrant a reason to rejoice just as the gentile with leprosy rejoiced knowing that Jesus saved him. Paul implies that we are now circumcised not by human hand, but through the divine hand of Jesus (Col. 2:11). He reconciling the two groups, notice that he says that from both groups (Jewish and non-Jewish people—GW translation) who were ἀμφότεροι "exactly the same" in value to be candidate for salvation God made one in Christ—this include the Jews who accepted Jesus as the Messiah and the converted ἔθνος "gentiles" or "nations." We are now citizens of God's family as the NKJ translates it.

These is conditional to only those who have γίνομαι "become" close to the Father through the holy blood of Jesus Christ and are filled with the same Spirit. Interesting enough the trinity is mention here as a reminder that this migratory reform was done by the Godhead in a co-operation. As a result every immigrant is now part "nations." We are now citizens of God's family as the NKJ translates it.

These is conditional to only those who have γίνομαι "become" close to the Father through the holy blood of Jesus Christ and are filled with the same Spirit. Interesting enough the trinity is mention here as a reminder that this migratory reform was done by the Godhead in a co-operation. As a result every immigrant is now part of the company of the redeemed of all ages beginning with Adam.[12]

"that formerly you, the Gentiles in the flesh" This is literally "nations" (ethnos). It refers to all peoples who are not of the line of Jacob.

In the OT the term "nations" (goim) was a derogatory way of referring to all non-jews.[13]

Grammar

mnemoneuo. This word is in present, active, imperative 2nd person, plural which means "to keep in mind" or "to keep thinking about." Commands a habitual action often expressing a note of urgency. It is also related to a long-term commitment and calls for the attitude or action to be one's continual way of life. The fact that it's in plural denotes that both male and female are included in the remembering.[14]

Interestingly enough, the LXX uses the term ***mimnesko*** in the fourth commandment which is akin to *mnemoneuo*; perhaps, Paul is trying to convey that the type of remembering is similar to the one God commanded on Exodus 20:8.

ethnos. Which probably comes from ethos [from where we get the word "ethics"], means "mass," "multitude," "host," and a human group.[15] In other words, ethnos means a nation with different customs, matters or different ethical values. When Jesus speaks about Gentiles, He uses this

---

[12] Hoehner, H. W. (1985). Ephesians. In J. F. Walvoord & R. B. Zuck (Eds.), *The Bible Knowledge Commentary: An Exposition of the Scriptures* (J. F. Walvoord & R. B. Zuck, Ed.) (Eph 2:19). Wheaton, IL: Victor Books.

[13] Utley, R. J. (1997). *Vol. Volume 8: Paul Bound, the Gospel Unbound: Letters from Prison (Colossians, Ephesians and Philemon, then later, Philippians)*. Study Guide Commentary Series (89). Marshall, TX: Bible Lessons International.

[14] "Greek Verbs Quick Reference," last modified December 01, 2011, accessed June 22, 2013, http://www.preceptaustin.org/new_page_40.htm.

term to refer to them (Mat. 6:32; Lk. 18:32).

apallotrioo. "To alienate" this is a perfect passive participle meaning "have been and continued to be excluded." In the OT this term referred to resident non-citizens with limited rights (aliens). The Gentiles, [as understood by the Jews,] had been and continued to be separated, alienated from the Covenant of YHWH.[16] We still use *Alien* today in our immigration system for immigrant converting to U.S citizens.

xenos. Words of the *xen-* stem can mean "foreign" or "strange" but also "guest." Strangeness produces mutual tension between natives and foreigners, but hospitality overcomes the tension and makes of the alien a friend."[17]

This is yet another term to refer to an immigrant, ***ginomai.*** to *cause to be ("gen" -erate)*, i.e. (reflex.) to *become (come into being)*, used with great latitude.[18] Notice that to become something it requires a sort of transformation. Paul alludes to the power of Christ to become a citizen.

The Church as a Welcoming and Integrative Community for Immigrants

Paul's exhortation in Ephesians 2 is more than a theological reminder—it is a call to identity and allegiance. His command to remember (μνημονεύετε) is a spiritual discipline, an act of continual reflection on transformation. Just as immigrants recall the land they left and the nation they have joined, believers are urged to never forget the transition from alienation to belonging—from ἀπαλλοτριόω ("noncitizen") to συμπολίτης ("fellow citizen"). This remembrance builds gratitude and loyalty, the same virtues that define a warrior's heart. Paul's emphasis on unity between Jews and Gentiles mirrors the cohesion required among soldiers who must fight as one body, regardless of background or origin.

---

[15] Kittel, G., Friedrich, G., & Bromiley, G. W. (1985). *Theological Dictionary of the New Testament* (201). Grand Rapids, MI: W.B. Eerdmans.

[16] Utley, 89.

[17] Kittel, 661.

[18] Strong, J. (2009). Vol. 1: *A Concise Dictionary of the Words in the Greek Testament and The Hebrew Bible* (20). Bellingham, WA: Logos Bible Software.

In this light, The Believing Warrior finds its grounding: service to God and service to one's nation both demand remembrance, discipline, and transformation. Paul reminds the Ephesians that citizenship in God's household was not achieved by merit but by divine cooperation—the Father's will, the Son's blood, and the Spirit's presence. Likewise, the soldier's transformation is not merely external but internal, shaped by commitment and obedience to a higher cause. The Greek γίνομαι ("to become") captures this process perfectly; it is not instant but formed through continual becoming. The believing warrior, once a foreigner to grace, now lives as both a heavenly and earthly citizen—guided by ethics (ἔθος), bound by covenant, and trained in gratitude.

Just as Christ reconciled divided peoples into one body, the faithful soldier embodies reconciliation through service, protecting peace while representing the values of the Kingdom. Remembering who we once were fuels our resolve to serve with humility, courage, and devotion. The believer's remembrance becomes the warrior's readiness—an active awareness that every act of service, whether spiritual or military, is a living testimony of the transformation made possible through Christ.

CHAPTER 2

The Pro's and Con's of Becoming a Service Member

*J*OINING THE MILITARY HAS ITS SHARE of advantages as well as disadvantages. Many advantages include tuition assistant, Montgomery GI Bill, job experience, Veteran Affairs home loans, travel around the world, health and dental care benefits, and self-pride. Within the disadvantages, there is the possibility of traumatic events, physical wear and tear, and even religious/spiritual disconnection.

Military Advantages

Educational

The biggest incentive it offered me back in 2003 when I joined was its educational benefits which guaranteed paying for college. The so-called GI Bill and Tuition Assistance (TA) was the two biggest determinants for why I joined. While stationed at Camp Lejeune, North Carolina, I was able to start college courses under TA which paid 100% of all my courses' costs. Later, when I got out of the service, I then used the GI Bill and paid for my Bachelors and Master's degree under the service-connected program aimed to help disable veterans to find a job. All my books, fees and equipment were paid for and even a small stipend helped keep my family afloat while completing these degrees. The bargain was ideal for what I needed. When the GI Bill ran out, additional benefits kicked in so I could complete my Master's degree in divinity.

Experience

Once I completed my schooling, it was time to find a job. At the time, I was doing additional training in Clinical Pastoral Education, a course about self-awareness, team cohesion and clinical mindfulness. I was learning how to recognize my own pain during critical moments with patients in order to be present with them. Three out of four units had passed when I got the call from a hospital CEO who somehow got my resume and wanted to interview me for an executive position. I was going to become the Director of an entire Pastoral Care Department; a coveted position that it is only given to someone with professional maturity. I got the job because I demonstrated confidence and my military background had a heavy determinant in persuading the hiring committee.

Veteran Affairs Home Loan

I accepted the job offer and began preparations to move to Texas where we lived for several years. Upon arrival to San Marcos, we were faced with the decision to rent or buy. At the time, the real estate market had lowered its interest rates and it made sense to purchase a home. We had no savings, barely any credit and no history of owning a home. All the odds were against us in purchasing a home. Everyone told us that we needed at least 20% down payment out of the total cost of a home which we did not have. It was not until then we disclosed that we were military to our realtor. She referred us to a lender who enjoyed working with VA loans because they were reliable. We purchased a new home and later a second one with zero down payment—all because of our service to the military.

Travel Around the World

The military has given me the opportunity to travel throughout multiple places. As part of one's military service, a service member enjoys relocation expenses coverage every change of stations. The amount covered ranges thousands of dollars and it is different from enlisted and officers. A family move from Texas to Maryland can range between $15-20k. The benefits even extend beyond active duty services where veterans are eligible to "space-available" free flights anywhere in the world there is a military air base. In addition, there are plenty of military friendly businesses who offer incredible deals to service members and some even offer giveaways. For example, Sea Worlds offer free entrance once a year up to six military family members. Golden Corral is well knowns for their free meals on Veteran's Day.

Health and Dental Care

Another benefit is the constant medical checkups service members must comply with. It is part of a service member routine to maintain medical readiness at all time. Unlike a civilian who has the option to schedule their own appointments, in the military there are designated individuals who maintain active excel spread sheets verifying that all members are up to date with their vaccines, dental cleanings, annual screenings, and even physical fitness. Since the Armed Forces benefits from a holistic wellness from every member, they maintain an annual physical readiness test which is mandatory. Given that this is an expectation, the military is dedicated in providing state of art gyms that allows service members to visit and exercise—all for free! The access to these gyms extend to family members. Therefore, a membership that would cost between $120-$200 for a family of two, is now free allowing service members the ability to save money.

Self-Pride

My father used to tell my brothers and I how much he wished to have joined the military. He was infatuated with the idea of discipline and the crisp uniforms. To his surprise, three out of four became military service members. Two joining the U.S. Marine Corps and one the U.S. Army. Our father is very proud to have children who wear the uniform and bring honor to a reality that was once a dream. The idea to bring a smile into the faces of my father and mother is priceless. This sense of pride carries throughout generations to the point that today our military service serves as an inspiration to our children and other extended family members. But more importantly, it brings pride to self. It creates a sense of fulfillment in that, as an immigrant, I was able to serve and I am deeply thankful for the opportunity.

Military Disadvantages

Traumatic Events

I will spare you from giving you the plethora of scientific literature on trauma linked to military service since the scope of this book is intended to encourage God-fearing individuals who want to join the military, but hesitate because of conviction. However, we cannot deny its challenges and possible experiences one may face. As a chaplain, I have seen countless cases of service members with a spectrum of traumatic stress. Some of them related to their service, but the majority related to their upbringing.

Military Sexual Trauma is one subject that has caught the attention of researchers. It is when a service member is exposed to assault, batter, or harassment with sexual nature. The exposure to a sexual trauma can indeed jeopardize one's mental health in unimaginable ways for the service members, their families and even the community.[1] This issue does not just affect females, as many may think, but also males. In a scientific study among Canadian Armed Force members, 44.6% of women experience military sexual trauma[2].

In the United States of American, males demonstrated having a high rate of sexual traumatic cases where 43,693 male veterans and 48,106 females were screened positive for military sexual trauma.[3]

In addition, Morally Injurious Events (MIEs) are understood as experiences that can undermine someone's basic sense of humanity and their comprehension or understanding of how the world operates.[4] MIEs are linked to violence, war related stressors (e.g., witnessing injured people, dead bodies, and fighting), natural disasters (e.g., floods, hurricanes, and earthquakes), and witnessing suicidal people or cases of rape, assault, or life-threatening accidents as predictors of trauma related distress, additional types of occupational stressors are challenging people's deeply held moral beliefs and values.[5]

We could also mention Post Traumatic Stress Disorder and depression have been associated with negative mental health outcomes for service members. This comes as a result of deployments to danger areas and the exposure to chaos, death and loss. Then, the service member returns home and is now faced with the adjusting to civilian life and a different cultural norm.[6]

As I have served for almost 14 years now, God has protected me from any of these traumatic experiences. I found courage in meeting countless of fellow God-fearing individual in the service who have been blessed with the same protection. We challenge the idea that just because these potential experiences exist in the military, it does not mean a service member will go through them. Likewise, in the world exist the possibility of similar exposure and repercussions. As a civilian pastor, I encountered

---

[1] Eckerlin et al., "Military Sexual Trauma," 34.
[2] Mota et al, "Prevalence and Correlates," 1.
[3] Eckerlin et al., 35.
[4] Currier et al., "Morally Injurious Experiences," 24-33.
[5] Kopacz et al., "A Preliminary Study," 1332.
[6] Hipes, Lucas, & Kleykamp, "Status and Stigma," 477.

a good share of individuals who were victims of abuse, sexual trauma, moral injury and post trauma.

Physical Wear and Tear

Indeed, physical wear and tear exist in the military in many ways, shapes or forms. Starting with recruitment, a person interested in joining the military must be able to at least show basic physical fitness. I remember when I first went to the Marine Corps' recruiter, I asked what I must do to join the service and he proceeded to a metal bar and asked me to do as many pull ups as I could. I did 8 at the time which he replied that I had done the minimum amount to join! During basic training, a recruit must demonstrate the ability to complete 13 weeks of rigorous physical challenges. This include lack of sleep, deprivation of food sometimes, and long training days.

Make no mistake, military does have a huge emphasis in staying active. Today, they have become more flexible in still holding service members accountable for outstanding physical fitness. Depending on the career a service member chooses, will also depend on how rigorous this requirement will be. For example, an administrative job may not be as physically demanding as someone in infantry who is always in the field training. I have to say though that regardless of military requirement, a God-fearing person should always maintain their bodies actively ready.

Even Religious/Spiritual Disconnection

It is true that the possibility of becoming disconnected from one's religious/spiritual walk is present. Again, this possibility is not greater or lesser just because someone joins the military. Yes, certain religious practices become more challenging than others. For example, Sabbath keepers may experience difficulties abstaining from work during a deployment or during an attack demanding all hands' reaction force. The sense of community is also different. I have to say that getting used to a new religious group can be challenging for some individuals who struggle with socialization.

One become more aware that a religious faith is not solely based on community, but merely on personal relationship with their own deity. In the example of Sabbath keeping, it comes through various forms where service members can ask a Commanding Officer to allow them with time to worship or engage in religious practices which can be granted or denied depending on its effect on the mission. For example, someone who desire to sacrifice an animal on board a Navy war vessel may find themselves getting a denial than someone who really values praying 5 times a day because of safety.

In Conclusion

The military, in all its complexity, mirrors the Christian journey. It offers rewards that build character, discipline, and purpose—but it also demands endurance, sacrifice, and faith under fire. Every advantage carries a cost, and every hardship presents a lesson in perseverance. Through this balance, we discover that the call to serve one's nation is not entirely separate from the call to serve God. Both require loyalty, courage, and submission to a mission greater than oneself.

The believing warrior recognizes that the battlefield is not only physical but spiritual. The same discipline that sustains a soldier in combat strengthens a believer in times of temptation. The same courage that faces danger in uniform equips the Christian to confront evil with conviction. Service, therefore, becomes sacred when it flows from faith—when a soldier's integrity, compassion, and obedience reflect the heart of God.

Paul's words to Timothy ring true for every believer in uniform: "Endure hardship as a good soldier of Jesus Christ" (2 Tim. 2:3). Military service is not simply a profession; it is a form of discipleship—a testing ground for faith lived out in motion. The believing warrior learns that weapons alone do not win battles; integrity does. Rank alone does not command respect; character does.

As we turn to the next chapter, we will explore how Scripture itself portrays the divine nature of service and warfare—how heavenly armies, spiritual armor, and righteous command all reveal that to fight for justice and peace is to walk in step with the Commander of Heaven's host. The believer who serves in uniform stands not in contradiction to faith but as its living expression: disciplined, courageous, and faithful to both God and country.

CHAPTER 3

The Opposition Says...

OPPOSITION TO MILITARY SERVICE from within Christian circles is not new. In fact, this conversation spans centuries and cuts across denominations, theological systems, and personal convictions. Many respected voices have passionately argued that participation in armed conflict is incompatible with the teachings of Jesus Christ. Their reasoning often centers around the sanctity of life, nonviolence, and a desire for believers to remain unstained by the systems of this world.

And to be fair, many of these objections come from a place of sincere moral concern. These are not enemies of faith—they are, in many cases, faithful followers of Christ trying to navigate the tension between kingdom ethics and earthly citizenship. Their convictions deserve to be heard. But they also deserve to be weighed—carefully, prayerfully, and biblically.

As someone who has walked both paths—the warrior's path and the disciple's walk—I believe we must look beyond surface-level assumptions and simplistic answers. This chapter explores the most common Christian objections to military service and considers whether those concerns hold up under the weight of Scripture and lived experience.

The Conservative Christian Perspective

Conservative Christians often uphold a theology of divine order and moral responsibility. They see government and military institutions as instruments God uses to restrain evil and preserve justice. Citing Romans 13:1–4, they argue that the governing authority "does not bear the sword in vain." For them, the "sword" represents legitimate power—the moral right to protect the innocent, uphold law, and punish wrongdoing.

Many conservative theologians believe that military service, when conducted with integrity and moral clarity, aligns with God's command to defend the oppressed. From this view, serving in the armed forces can be a sacred duty—a tangible expression of love for one's neighbor. To them, refusing to protect others from harm when capable of doing so is not peace—it's neglect.

In this perspective, the Christian soldier is not an aggressor but a guardian. War, though tragic, is sometimes the necessary response to injustice. Thinkers like Reinhold Niebuhr argued that pacifism, though idealistic, can become passive in the face of evil. As he wrote, "To defend democracy with arms is not to betray the gospel—it is to resist the powers that destroy it." For many conservative Christians, the soldier's uniform and the believer's armor of faith are not contradictory but complementary.

The Liberal Christian Perspective

Liberal and progressive Christians, on the other hand, emphasize nonviolence, reconciliation, and the radical ethic of Jesus' kingdom. They look to the Sermon on the Mount—"Blessed are the peacemakers" (Matthew 5:9) and "Turn the other cheek" (Matthew 5:39)—as the blueprint for Christian ethics. They argue that violence, no matter how justified, perpetuates the cycle of sin and dehumanization.

For many within this camp, the teachings and example of Jesus represent a complete departure from the logic of warfare. They point to Christ's refusal to summon angels to defend Himself at His arrest, His silence before Pilate, and His command to love one's enemies as evidence that the Christian's weapon is compassion, not combat. Early church fathers like Tertullian and Origen shared this conviction, teaching that believers should not participate in the taking of life—even for the state.

Modern voices continue that tradition. Many liberal theologians and denominations oppose war, viewing military institutions as extensions of worldly power that distract from the mission of God's kingdom. They advocate for service in noncombatant roles, humanitarian missions, or peacekeeping forces as expressions of "fighting for peace" without resorting to violence. Their question is not whether evil exists—it is whether violence ever truly conquers it.

Common Ground: The Moral Burden of Violence

Despite their differences, both conservative and liberal Christians share one profound truth: violence is never morally neutral. Both sides acknowledge that taking life—even in defense—carries a spiritual cost. The question, then, is not simply whether Christians can serve, but whether they can do so righteously.

War has consequences that echo beyond the battlefield. Even justified conflicts can leave moral scars. Conservative Christians who affirm "just war" principles emphasize the need for restraint, lawful authority, and right intention. Liberal Christians, while rejecting violence, remind us of the importance of compassion and repentance. Together, their perspectives remind the believing warrior that faith must never be overshadowed by force.

The Case Against Military Service

Three primary objections tend to arise from Christians opposed to military involvement:

- The ethical dilemma of bearing arms and possibly taking life
- The concern that military life leads to spiritual compromise
- The belief that military service fundamentally contradicts the teachings of Jesus

These arguments are usually framed with scripture in mind—verses like *"Thou shalt not kill"* (Exodus 20:13) and Jesus' call to *"turn the other cheek"* (Matthew 5:39). The reasoning follows a certain logic: If Jesus is the Prince of Peace, how could His followers be agents of war?

The Burden of Bearing Arms

The image of a Christian holding a weapon is unsettling for many. It raises hard questions: Can a believer who follows the Prince of Peace also carry a rifle? Can someone serve both the Kingdom of God and a nation-state that wages war?

Critics argue that carrying a weapon—even with noble intentions—inevitably puts a believer at odds with Jesus' teachings. They fear that military service requires a level of moral compromise that no Christian should be forced to make, especially if one's role involves using lethal force. The fear is not just about external violence—it's also about the internal erosion of one's conscience.

And the concern is understandable. War is violent. Combat changes people. But does this mean all service is incompatible with faith? Or is there another way to think about service—one that acknowledges both the weight of violence and the possibility of righteous intervention?

A Secular Institution and a Sacred Faith

Another argument leveled against military service is that the environment itself is spiritually corrosive. The military is often perceived as harsh, hierarchical, and worldly. Critics point to the prevalence of profanity, immorality, and a mission-driven culture that may seem to leave little room for spiritual nourishment or community.

It's true that the military can be spiritually lonely. The rhythms of worship, fellowship, and church life are often disrupted by deployments, missions, and rotations. But these realities are not unlike those faced by missionaries, humanitarian workers, or anyone who ventures beyond the comfort zone of church walls. The real question is: **Does a hostile environment disqualify a Christian from entering it—or does it make their presence more necessary?**

Biblical and Theological Concerns

From a theological standpoint, some argue that military service is fundamentally at odds with Christian identity. We are, after all, citizens of another kingdom. Paul writes, *"Our struggle is not against flesh and blood"* (Ephesians 6:12), reminding believers that our true warfare is spiritual. Christ's followers are to be peacemakers, not participants in systems of destruction.

But is this the full picture?

The Bible is filled with godly men and women who were called into contexts of conflict—sometimes violent, sometimes redemptive. The spiritual realm is not a safe place; it is a battleground. To reduce Christianity to passive resistance ignores the active, confrontational, and even warrior-like imagery found throughout scripture.

Peter and the Sword: A Misunderstood Moment

One of the most frequently cited passages against military service is found in the Garden of Gethsemane. Peter, attempting to protect Jesus, draws his sword and strikes the ear of Malchus, a servant of the high priest. Jesus' response is swift and sharp: "Put your sword away! Shall I not drink the cup the Father has given me?" (John 18:11).

At first glance, it appears Jesus is rebuking violence. But a closer reading reveals something deeper. Jesus never told Peter not to carry a sword. In fact, in Luke 22:36, Jesus tells His disciples to arm themselves. The issue is not the presence of the sword—it's the timing and purpose behind its use.

Peter misunderstood the moment. He tried to fight a spiritual battle with physical force. Jesus wasn't condemning defense or protection—He was clarifying His mission. The cross could not be avoided through violence. Redemption required surrender.

This moment does not invalidate service; it calls us to discernment. There are times when fighting is wrong. And there are times when it is necessary—not for revenge or conquest, but for justice and protection.

Warriors in Scripture

The idea that military service is inherently sinful or incompatible with faith is not supported by the whole of Scripture. On the contrary, many of God's chosen instruments were men and women who fought battles—physically and spiritually—for the sake of righteousness.

- **David,** a man after God's own heart, was a warrior-king who led armies with courage and humility. His psalms are filled with cries for divine protection in battle.
- **The Roman centurion** in Matthew 8 was a career military officer, yet Jesus marveled at his faith—declaring it greater than any He had found in Israel.
- **Michael the archangel** leads heavenly armies against the forces of darkness (Revelation 12:7). Even heaven wages war—because evil must be confronted.

Reconciling Faith and Military Service

We must ask ourselves: If the battlefield is one of the darkest places on earth, why wouldn't God send His people there?
Salt belongs in what is decaying. Light belongs where there is darkness. A Christian who joins the military does not betray the gospel—they may, in fact, become its most needed ambassador in uniform.

Yes, there are ethical complexities. Yes, the risks are real. But the answer is not to retreat—it is to engage with wisdom, conviction, and an unshakable connection to the Vine (John 15:5). Military service is not a detour from discipleship—it can be one of its most intense expressions.

A Calling, Not a Compromise

Some will still disagree. And that's okay. Conscientious objection has its place in Christian history. But for others—those who feel a stirring, a burden to serve—their calling should not be dismissed or shamed. Not every soldier is driven by nationalism. Some are compelled by conviction, by a desire to protect, to heal, to carry burdens with their brothers and sisters.

Jesus prayed, "My prayer is not that you take them out of the world but that you protect them from the evil one" (John 17:15).

This prayer is as relevant to believers in uniform as it is to believers in pulpits. God does not call everyone to the same battlefield—but He does call each of us to follow Him with integrity wherever we are sent.

In Conclusion

The opposition to military service among Christians stems from genuine concern, but it must be weighed against the fullness of biblical truth. Military life is not easy, nor is it always understood by those on the outside. But for the believer who enters it prayerfully, humbly, and anchored in Christ—it can become a profound place of transformation.

Military service, when surrendered to God, can become a vessel of grace. Not in spite of the uniform, but through it.

I know this, not because I read it in a book, but because I lived it.

CHAPTER 4

God May Have a Plan for You

THE QUESTION THAT ECHOES in the heart of many young men and women considering military service is a deep and personal one: What if God is calling me to serve? It's not just a logistical or career question—it's a spiritual one. It's a question that speaks to purpose, calling, identity, and faith. I remember wrestling with that question myself. When I joined the Marine Corps in 2003, I didn't yet know how to answer it. I didn't realize that behind my search for discipline, belonging, and a fresh start was a divine hand guiding my steps.

In hindsight, I see more clearly now that God had a plan. Not just a general plan, but a personal one—tailored to my fears, my background, my strengths, and even my failures. My military journey was not a detour from God's will; it was the very road He used to bring me to Himself and into ministry. And I believe the same may be true for you.

A Journey of Faith Begins

I was twenty-one, an immigrant still adjusting to a new country, still learning the language and customs of the United States. I didn't grow up dreaming of the military. I certainly didn't see myself as someone who would one day wear the uniform of the United States Marine Corps. But life has a way of leading us to unexpected places, especially when God is working behind the scenes.

When I enlisted, I was looking for structure and purpose. I wasn't thinking about theology or ministry—I was thinking about survival. I needed a way forward. But God had more in mind than simply helping me find direction; He wanted to reshape my identity.

Two years into my service, during a quiet evening at Camp Lejeune, something stirred in me. I wasn't in a chapel. I wasn't listening to a sermon. I was just alone with my Bible. I opened to Jeremiah 1:5, and the words leapt off the page: "Before I formed you in the womb, I knew you. Before you were born, I set you apart. I appointed you as a prophet to the nations."

I had read that verse before, but that night it felt personal. Could it be that God had seen me long before I saw Him? Could it be that my military service wasn't a detour—but preparation?

When Calling and Obedience Intersect

That moment sparked a journey of discernment. At first, I felt unqualified. I was a Marine, an immigrant, and someone with no formal religious training. Could God really be calling someone like me into ministry? It seemed unlikely, even foolish. But calling rarely waits until you feel ready. It calls you as you are—and shapes you as you go.

In 2005, I could no longer ignore it. The pull toward ministry was undeniable. And rather than asking me to abandon my military path, God was inviting me to build upon it. My training, discipline, teamwork, resilience—all of it would become tools for a much greater mission: to shepherd souls and proclaim His Word.

That transition didn't happen overnight. Like many who wrestle with God's will, I had to walk through uncertainty, doubt, and fear.

But step by step, God led me forward—through Scripture, prayer, and community—and confirmed that His hand had been on my life all along.

Blessings Along the Way

Obedience to God's call is often the gateway to blessings we could never have anticipated. It was during this season of transition that I met the woman who would become my wife—a Puerto Rican woman with fierce faith and an unshakable heart. She has been my anchor ever since. We married soon after, and together we built a life centered around faith, love, and service.

In 2007, we were blessed with our daughter—a beautiful reminder of God's generosity and grace. Parenthood deepened my understanding of what it means to lead, to serve, and to love unconditionally. Every diaper changed, every prayer whispered at night, every moment of joy and exhaustion—they all became part of my spiritual formation.

With a growing family and a burning sense of purpose, I pursued a bachelor's degree in theology, followed by a Master of Divinity, and eventually, a doctorate in philosophy. That journey was not easy. Balancing military responsibilities, school, ministry, and fatherhood was a test of perseverance. But in every challenge, God's provision was clear. He didn't call me to something He would not also equip me for.

The Military as God's Refining Furnace

Looking back, I see that the military didn't pull me away from God—it brought me closer. The Marine Corps became a refining furnace for my character. It taught me endurance, accountability, humility, and self-discipline. It put me in situations where prayer was not optional, but vital. It exposed my weaknesses, tested my limits, and strengthened my trust in God.

For those who fear that the military might be too secular or too distant from faith, I want to offer this perspective: the military can be one of the most fertile grounds for spiritual growth. Where there is risk, there is dependence. Where there is suffering, there is often clarity. And where there is purpose, there is often calling.

Overcoming Fear and Trusting God's Purpose

Fear has a way of standing in the doorway of our calling. It whispers that we're not good enough, not holy enough, not strong enough. But those are lies. As Paul reminded Timothy, "God has not given us a spirit of fear, but of power, love, and a sound mind" (2 Timothy 1:7). If God is prompting your heart to consider military service—not as an escape, but as a calling—then you can walk that path boldly, knowing He walks with you.

You don't need to see the full picture. You don't need all the answers. You just need enough faith to take the next step. And often, that's how calling works: one obedient step at a time.

The Call You Don't Recognize

Many people believe God's call must sound like a trumpet—loud, clear, unmistakable. But my experience has taught me that sometimes it comes quietly, disguised as confusion or necessity. I was 21, an immigrant from Central America, barely fluent in English. I had survived culture shock, financial hardship, and spiritual uncertainty. I didn't join the Marines because of a vision from heaven. I joined because I didn't know what else to do. The world around me offered little guidance, and the future was a fog of uncertainty. I needed a way forward. And the Marine Corps, of all places, opened its doors.

I wasn't aware that this was divine direction. I had no theology of vocation. I didn't understand calling. But even though I was unaware, God was not absent. He was working behind the scenes, orchestrating circumstances, softening my heart, preparing divine appointments, and setting up lessons I could never have learned in a seminary or classroom.

When the Uniform Meets the Altar

I learned to polish my boots before I learned to exegete a passage. I stood at attention before I ever stood behind a pulpit. And I raised my right hand to defend the Constitution before I ever lifted both hands in worship as a pastor. The military, far from being the antithesis of faith, became the arena where God forged the character that ministry would demand of me. And this realization—this fusion between military discipline and spiritual devotion—shook me to my core.

God used my time in the Marines as a divine boot camp for my soul. The strict discipline, the training in endurance, the value of integrity, and the painful breaking of self-will were all lessons I didn't know I needed. It was there that I learned how to submit to authority, not just to commanding officers but to the King of Kings. It was there that I began to understand what it meant to serve others, to endure hardship, and to push beyond my comfort zone. I came to realize that basic training was not just breaking my body—it was breaking my pride. And only through that breaking could God begin to build me up as a man of purpose.

Your Place in a Bigger Story

One of the most liberating truths I've come to accept is that we are part of a much bigger story. You may think your decision to enlist is just about benefits, school, or structure. And perhaps those things matter. But what if your story is connected to someone else's salvation? What if your assignment is not just to fire a rifle or serve a command—but to plant seeds of the gospel in hearts you would never meet otherwise?

Jesus said in Matthew 5:13-14, "You are the salt of the earth... You are the light of the world." Where is salt needed? In tasteless places. Where is light needed? In darkness. If you only stay in sanitized environments surrounded by fellow believers, how can your faith make an impact? The military places believers in some of the most spiritually desolate environments—places where vulgarity, pride, trauma, and despair run rampant. And it is in those very places that your light is most needed.

The idea that God only calls people into pastoral ministry or missionary work overseas is a tragic misunderstanding of Scripture. Throughout the Bible, God sends His people into the heart of conflict—not to retreat, but to engage. Think of Joseph, sent into Egypt. Think of Daniel, promoted in Babylon. Think of Esther, raised for "such a time as this" in the Persian palace. None of them were in traditionally "holy" places. And yet God placed them there as agents of transformation.

When God Sends You into Babylon

The military may feel like Babylon—a place of competing loyalties, of harsh rules, and a culture that doesn't always reflect the values of the Kingdom. But remember this: God never needed a perfect environment to do perfect work. In fact, it is often in the "Babylons" of life that God raises up His most faithful servants. In Jeremiah 29, God gives a shocking command to the Jewish exiles: "Build houses and settle down... Seek the peace and prosperity of the city to which I have carried you into exile. Pray to the Lord for it" (Jer. 29:5-7). He didn't tell them to escape Babylon. He told them to bless it.

That was a paradigm shift for me. For so long, I thought faith was about escaping difficult environments—running toward comfort and away from chaos. But Scripture teaches us that sometimes God sends us right into the heart of chaos, not to be consumed by it, but to influence it.

In the same way, the military may not look like a spiritual mission field, but it is. It's a unique place where people are broken, searching, open, and willing to talk about life and death in ways they wouldn't in ordinary civilian life. That's why your presence there matters. You may be the only living Bible someone ever reads. You may be the only preacher they'll listen to. You may be the friend who listens when others don't.

A Theology of Vocation and Calling

Too often, believers speak of "calling" only in church-related terms. But Scripture doesn't make such a distinction. Colossians 3:23 reminds us,

"Whatever you do, work at it with all your heart, as working for the Lord, not for human masters." Whether you're a pastor or a pilot, a chaplain or a corporal, your service can be sacred if your heart is surrendered to God.

Martin Luther once said, "The maid who sweeps her kitchen is doing the will of God just as much as the monk who prays—if she does it for the glory of God." The same principle applies to soldiers, airmen, sailors, and marines. The question is not the nature of the work—but the posture of the heart. Could it be that the military is not just a career, but a calling? Could it be that God is positioning you to be a Daniel in your barracks, an Esther in your chain of command, a Joseph in your deployment?

There Is No "Secular" for the Surrendered

One of the biggest spiritual lies is the separation between "sacred" and "secular." Too many Christians think God is only interested in what happens within church walls—sermons, worship, and prayer. But if God only called us to serve in churches, who would reach the streets? Who would reach the military bases? Who would shine in boardrooms, classrooms, or conflict zones?

The truth is this: for the fully surrendered life, there is no such thing as secular work. Everything becomes sacred when offered in obedience. Whether you're guarding a post, attending boot camp, or flying a mission, your life can be an altar. Your obedience, your attitude, your integrity—these are offerings to God. Romans 12:1 urges us to "offer your bodies as a living sacrifice, holy and pleasing to God—this is your true and proper worship." That means worship isn't just a Sunday activity—it's a lifestyle.
God doesn't need you to wear a clerical collar to use you. He just needs your "yes."

Salt for the Tasteless

Jesus calls His followers "the salt of the earth" (Matthew 5:13). Salt is more than seasoning; it's preservation. In biblical times, before refrigeration existed, salt kept food from rotting. It was rubbed into meat to preserve it from decay. That image is powerful. Salt doesn't do its job by staying in the jar. It only works when it's rubbed into raw, decaying flesh. In the same way, the presence of believers in corrupt environments doesn't contaminate them—it preserves them.

Could it be that God wants to send you into the "raw" places of life? Into units plagued by vulgarity, into deployments heavy with despair, into communities where violence and hatred reign? And not to judge them—but to preserve them? Your presence can hold back moral and spiritual decay. Your life can interrupt a cycle. Your courage can prevent a suicide. Your character can restore honor. God can use you to protect lives in more ways than one.

Salt also creates thirst. A life lived in alignment with Christ causes others to thirst for what you have. I remember during my deployment, one of my fellow Marines approached me after a grueling day of training. He didn't ask for a theological debate or a church invitation. He just said, "There's something different about you. Can we talk?" He was wrestling with guilt from home, battling depression, and questioning his purpose. And in that moment—because of how I lived, not just what I said—God opened a door for the gospel.

Your saltiness matters. Your presence matters. The way you treat your commanding officer, the way you carry your weapon, the way you show compassion to your brothers and sisters in arms—these are sermons louder than words.

The Power of Purpose in Unlikely Places

It's easy to feel like the military is too rough, too secular, too distant from the church to be used by God. But let me ask you this: where else do you find young people voluntarily submitting themselves to life-altering discipline, self-denial, and loyalty to a mission greater than themselves? In some ways, the military offers a clearer picture of the Christian life than many churches do.

In the Marines, we were taught values like honor, courage, and commitment. We were trained to endure hardship, to respect authority, and to protect our own—even at the cost of our lives. These values don't contradict Scripture; they reflect it. Jesus said in John 15:13, "Greater love has no one than this: to lay down one's life for one's friends." Soldiers live this verse every day.

What if God could use the military as a crucible to refine your character? What if boot camp becomes your desert, like Moses had? What if deployments become your mission field, like Paul's journeys? What if your service becomes your sanctification?

Let's stop limiting God to religious spaces. He's the God of the universe, not just the church. He called fishermen and tax collectors. He met shepherds in fields and wise men in palaces. He can meet you in your barracks, on the battlefield, or in a briefing room.

Your Testimony is Your Weapon

One of the most powerful tools a Christian has in any environment is their testimony. You don't have to be a theologian. You don't need a seminary degree. You just need to tell your story—what God has done for you, how He changed your life, what He brought you out of.

Revelation 12:11 says, "They triumphed over him [Satan] by the blood of the Lamb and by the word of their testimony." Your story has power. And in the military, stories matter. When you're serving side-by-side with someone through exhaustion, fear, or loss, the walls come down. People are more open than you realize. And when they see that you have peace in chaos, strength in weakness, and hope in darkness, they'll want to know why.

Be ready to tell them. Be ready to say, "I'm not perfect, but I know the One who is. I've been broken, but I've been healed. I've seen darkness, but I walk in the light. And I believe He has a plan for you, too."

Don't underestimate how God can use your story. Don't underestimate how many are watching you, even silently. They may never come to church, but they'll remember the way you forgave that insult. They'll remember how you stayed calm under pressure. They'll remember how you prayed in secret, how you encouraged the weary, how you stayed clean when others compromised. And one day, they'll come to you—and ask why. And on that day, your testimony becomes your weapon.

Ministry Doesn't Always Wear a Robe

One of the greatest misconceptions in the Church is the idea that ministry is confined to a pulpit. That unless you're preaching, teaching, or leading worship, you're not really "doing ministry." But Scripture paints a much broader picture of God's calling. Ministry isn't a title—it's a lifestyle. It's serving others with love, truth, and humility wherever God places you.

In the military, your pulpit might be a bunk bed in a barracks.

Your congregation could be your squad. Your sermons might be delivered through actions, not words—through discipline, respect, and integrity. I've seen more gospel opportunities around a fire watch shift than I did in some church pews. Why? Because authenticity thrives in raw spaces. When men and women are stripped of comfort, rank, and routine, they become more open to real conversations—conversations about death, fear, identity, and eternity.

You may be the only Bible someone ever reads. In moments of exhaustion, doubt, or despair, your consistent example can preach louder than a thousand sermons. Ministry is being present when others walk away. It's speaking peace in a storm, showing kindness in a harsh world, and offering hope when there seems to be none. It's being an ambassador for Christ in boots and camouflage.

This is the heart of incarnational ministry—God taking flesh in real places through real people. He did it in Jesus. He wants to do it in you.

How Do You Know If It's God's Plan?

This is the question that haunts many young people: How can I know if the military is God's will for me?

There's no one-size-fits-all answer, but there are biblical principles to help guide your discernment:

1. Are you seeking God in prayer? God is not silent. He speaks through His Word, His Spirit, and His people. If you're considering military service, begin with consistent prayer. Ask Him to close the doors that aren't from Him and open the ones that are. (James 1:5).

2. Do others confirm your calling? Wise counsel matters. Talk to spiritual mentors, pastors, and mature believers. Ask them to pray with you. Often, God uses others to confirm what He's already been stirring in your heart. (Proverbs 11:14)

3. Are your motives aligned with His purposes? Ask yourself honestly: Why am I joining? If it's for revenge, pride, or money alone, you may need to pause. But if you're sensing a deeper calling to serve, protect, grow, and witness—God may very well be in it. (1 Samuel 16:7)

4. Does it bring peace—not ease, but peace? God's plan doesn't always lead to comfort, but it always brings peace. There's a difference. Jonah had no peace on a boat in the wrong direction, but Paul had peace in prison. Let the peace of Christ rule in your heart. (Colossians 3:15)

God's plans often stretch us. They rarely make sense in the beginning. But they're always for our good and His glory.

When the Military Becomes the Mission Field

Every believer is a missionary, whether they're on foreign soil or in their hometown. But the military presents a unique mission field—one filled with young people from every state, every background, every faith (or lack thereof). You'll rub shoulders with people you might never meet otherwise. And many are searching—hungry for meaning, for purpose, for something real.

You don't have to go overseas to fulfill the Great Commission (Matthew 28:19–20). Sometimes God brings the nations to you through your platoon. You may sit beside a Muslim during a flight, bunk next to an atheist during training, or share a meal with someone who's never heard the gospel in their entire life. What will you do with that opportunity?

I've had moments in the military where I questioned whether I was making any spiritual impact. But God reminded me—it's not about numbers. It's about faithfulness. A seed planted in silence can one day bloom in boldness. Your job isn't to convert—it's to represent. Be the aroma of Christ (2 Corinthians 2:15) and let the Spirit do the rest.

Don't Disqualify Yourself

I once believed that my background disqualified me from ministry. I was an immigrant. English wasn't my first language. I didn't grow up in the church. I carried pain, doubt, and pride. I saw my flaws more than my potential.

But God specializes in using the unexpected. He chose Moses—a stutterer—to speak before Pharaoh. He chose David—a shepherd boy—to defeat giants. He chose Esther—an orphaned exile—to save a nation. He chose fishermen, tax collectors, and even zealots to become His disciples. Why wouldn't He choose a Marine?

Whatever your background, your failures, or your fears—God can use you. In fact, He wants to use you. Don't let your past keep you from your purpose.

As Paul wrote in 1 Corinthians 1:27–29:
"But God chose the foolish things of the world to shame the wise; God chose the weak things of the world to shame the strong… so that no one may boast before him."

Your weakness may be the very vessel through which God displays His strength.

From Camouflage to Calling

The journey from being a soldier to being a servant of the gospel may seem like two different paths, but they are often deeply intertwined. When I first stepped into boot camp, I didn't imagine God would use those same steps to guide me into ministry. Yet, time after time, I've seen how military experiences mirrored spiritual realities.

- **Discipline** became devotion.
- **Obedience to command** became obedience to God.
- **Teamwork and brotherhood** prepared me for pastoring a church.
- **Mission focus** taught me the urgency of the gospel.

The military taught me how to carry responsibility, how to fight through adversity, and how to protect those under my care—lessons that would later become the core of my pastoral calling. God used the uniform to prepare the vessel.

I often think of Joseph in Genesis. He was sold into slavery, imprisoned, and forgotten. Yet, each chapter of his life was a divine setup for the purpose God had for him: to save many lives. He told his brothers, "You intended to harm me, but God intended it for good" (Genesis 50:20).

You may not understand how it all fits together right now. But God is a master of weaving together broken pieces into beautiful outcomes.

God's Voice in Unexpected Places

Some of my deepest spiritual insights came not in a church, but in the quiet moments after patrols, the stillness of a watch post, or the silence in a tent under the stars. The military strips away distractions. It forces you to confront yourself. And in that raw honesty, God often speaks.

- He speaks in your exhaustion, reminding you that His strength is made perfect in weakness (2 Corinthians 12:9).
- He speaks in your fears, whispering that He has not given you a spirit of fear but of power, love, and a sound mind (2 Tim. 1:7).
- He speaks in your loneliness, affirming, "I will never leave you nor forsake you" (Hebrews 13:5).
- He speaks in your doubts, assuring you that "He who began a good work in you will carry it on to completion" (Philippians 1:6).

Maybe you've been waiting for God to show up in stained glass and organ hymns, but He's been walking beside you in boots and combat gear the whole time. God's presence is not bound to buildings. His voice is not limited to pulpits. He can speak in the field just as clearly as He speaks from a mountaintop.

Salt and Light in a Tasteless World

In Matthew 5:13–14, Jesus calls His followers "the salt of the earth" and "the light of the world." These words aren't just poetic—they're purposeful. Salt preserves and purifies. Light exposes and guides. Both require presence. You can't flavor food unless you're in the dish. You can't light a room unless you're in the darkness.

This is why your presence in the military matters. The barracks need salt. The battlefield needs light. Not everyone in uniform knows God, but they can know someone who does—you.

I remember one late night when a fellow Marine knocked on my door. He wasn't a Christian. He barely knew me. But he had heard from someone else that I prayed. That night, with tears in his eyes, he asked if I could pray for him. He was tired. Angry. Lost. He didn't want a sermon. He just needed a friend with faith.

In that moment, I wasn't a preacher—I was a vessel. And that was enough.

You may never stand behind a pulpit, but your life can preach. Your attitude can heal. Your example can challenge. You don't have to be perfect. You just have to be present.

Making Peace with a Warrior's Calling

There's a tension in being a Christian and a soldier. Some believers feel guilty for putting on a uniform. Others wrestle with the ethical implications of war and violence. These are valid concerns—and ones we must take seriously.

But remember this: God never called us to be careless. He called us to be faithful.

The Bible is filled with warriors whom God used mightily. Joshua. Gideon. Deborah. David. These were people who knew how to fight, but also how to worship. They were not perfect, but they were available.

Being in the military does not mean you are choosing violence. It means you are standing in the gap—often in places of chaos—to bring order, peace, and protection. The heart of a godly soldier is not aggression—it's service.

And isn't that what Jesus modeled? He came to serve, not to be served (Mark 10:45). He confronted injustice. He protected the vulnerable. He laid down His life for His friends.

You can be a warrior and a worshiper. The two are not mutually exclusive—they are biblically compatible.

Don't Wait for the Perfect Moment

Too many people delay obedience waiting for the "perfect" time. But here's a truth: the perfect time rarely comes. God doesn't always wait for us to feel ready—He moves when we're willing.

If I had waited until I had all the answers, I never would have joined the military. If I had waited until I felt qualified, I never would have gone to seminary. If I had waited until I felt holy, I never would have preached.

Obedience often precedes clarity. So, if you're standing at the crossroads, wondering if the military could be part of God's plan for your life—don't ignore the tug on your heart. Pray. Seek counsel.

Examine your motives. And then, trust that God's path may look different from what you expected, but it will always lead you closer to Him.

You May Be the Answer to Someone's Prayer

One of the most sobering realizations I've had is that sometimes you are the miracle someone else is praying for.

We often pray for God to intervene in a broken world, but what if His answer is to send you? Not because you're perfect, but because you're available. Not because you have all the answers, but because you've walked the hard road and still believe. Not because you're a saint, but because you've been through fire and still carry light.

When Jesus said, "You are the light of the world," He wasn't only speaking to pastors, missionaries, or seminary graduates. He was speaking to fishermen, tax collectors, ordinary people. He was saying, "The world is dark—and I'm putting My light in you."

That light doesn't go dim when you wear a uniform. In fact, it might shine brighter.

In your unit, someone is struggling with depression. In your barracks, someone is questioning their worth. On deployment, someone is asking if God even exists. In your circle, someone is praying for peace, for strength, for hope. And your faith, your presence, your courage may be the exact instrument God uses to speak to them.

Don't underestimate what God can do through you in the everyday spaces of service.

God Prepares in Hidden Places

Before David was king, he was a shepherd. Before Moses was a prophet, he was a fugitive. Before Esther was a queen, she was an orphan. Before Jesus launched His public ministry, He spent 30 years in obscurity. The military may be your "wilderness," your "training ground," your "hidden season." But don't mistake hiddenness for irrelevance. God often does His best work when no one else is watching.

- It's in the fire that gold is refined.
- It's in the desert that character is shaped.
- It's in the silence that God speaks the loudest.

If you feel unnoticed, misunderstood, or uncertain about the future, take heart. God may be preparing you for something greater than you can comprehend. You're not being forgotten—you're being forged.

Called to Serve, Not to Settle

One of the biggest lies you can believe is that your faith must be confined to the "religious" sphere. That if you're serious about serving God, you must become a pastor, a missionary, or a theologian.

That is simply not true.

Scripture is filled with examples of people who served God outside the temple:

- Nehemiah was a cupbearer—turned city rebuilder.
- Daniel was a government official—who influenced kings.
- Lydia was a businesswoman—who opened her home to the early church.
- The centurion was a Roman soldier—whose faith amazed Jesus.

God doesn't call everyone to leave the "secular" for the "spiritual." Sometimes, He sends you into the secular world so that the spiritual can be revealed there. The military is not outside of God's reach—it can be at the center of His mission for your life.

If God is Calling, Will You Go?

Maybe you're standing where I once stood: young, uncertain, unsure if God could ever use someone like you.

I'm here to tell you He can. He does. And He will—if you let Him.

Maybe military service is just a stepping stone. Maybe it's a training ground. Or maybe it's the very battlefield where your calling will come to life. Either way, it is not outside of God's jurisdiction. It may be the very place where He plans to meet you, shape you, and send you out with a message only you can deliver.

Don't dismiss the possibility that God's plan for your life includes boots, camo, and the call to serve.

Closing Words to the Reader

If you've read this far, maybe something deep inside you is stirring. Maybe you feel the pull to serve—not just your country, but your God. Maybe you're wondering if this chapter of your life could be a chapter in His story.

Here's what I want you to take away:

- God's plan is not limited by geography, career, or circumstance.
- He can use a uniform just as powerfully as a robe.
- He can speak through orders and oaths as clearly as He speaks through scripture.
- And He can take someone who feels unqualified and raise them to influence nations.

My journey through the military wasn't just about service—it was about surrender. And in surrender, I found calling. I found purpose. I found God. He may be calling you too—not just to wear a uniform, but to be a warrior of light in a world that desperately needs hope.

Will you answer?

CHAPTER 5

To Kill or Not to Kill... That is the Question

THE ACT OF TAKING A LIFE raises profound theological, moral, and ethical questions, especially for believers serving in the military. The tension between God's commandment—"Thou shalt not kill" (Exodus 20:13)—and the realities of warfare forces Christians to wrestle with how their duty to protect others aligns with divine will. For some, this conflict creates deep moral uncertainty: can one faithfully follow Christ while bearing arms?

This chapter examines that question by exploring a theological framework for understanding killing in Scripture, analyzing the Hebrew of Exodus 20:13, and distinguishing between God's perfect ethic and humanity's fallen one. It also considers the believer's opportunity to give life even in environments that deal with death. Throughout, the purpose is not to justify violence but to understand its moral weight in light of divine justice and mercy.

Faithful soldiers must not approach war lightly. Yet neither should they bear false guilt when fulfilling lawful duty. To kill or not to kill is not merely a question of physical action—it is a question of love, motive, and moral alignment with the God who values every human life.

Killing in the Old Testament: A Theological Framework

The act of taking a human life is one of the most sobering moral questions in all of Scripture. It sits at the intersection of divine command, human fallenness, and ethical responsibility. The Old Testament does not shy away from this complexity. Instead, it confronts it directly, weaving together narratives of war, justice, and mercy in a manner that both shocks and instructs the modern reader. When examined closely, it becomes clear that the God of the Old Testament neither glorifies violence nor ignores its necessity within a fallen world. Rather, He limits it, contextualizes it, and redeems it through justice.

At the center of this moral tension stands the Sixth Commandment: *"You shall not kill"* (Exodus 20:13, KJV). For many believers, this verse has become the cornerstone argument against Christian participation in war or any act of lethal force. Yet the original Hebrew sheds crucial light on its meaning and scope. The command uses the word רָצַח (ratsach), which, when carefully studied, translates more precisely as *murder* rather than the general term kill. Murder refers to the *intentional, unlawful, or vindictive* taking of innocent life. The ancient Hebrew had other words for killing in general—הָרַג (harag), meaning "to slay," or שָׁחַט (shachat), "to slaughter." The choice of *ratsach* was deliberate, distinguishing morally unjustifiable killing from acts permitted under divine authority.

This nuance shows that the commandment was not a blanket prohibition of all killing but a safeguard against the desecration of human life. Israel's theocratic system included legitimate occasions where lethal force was sanctioned: judicial execution (Deut. 19:11–13), national defense (Deut. 20), and divinely ordained warfare. These were not arbitrary acts of aggression but expressions of divine justice against profound wickedness. In fact, the Torah includes extensive guidelines meant to restrict violence and protect both the innocent and the environment even in wartime.

Deuteronomy 20, for instance, outlines a theology of restraint. Before attacking a city, Israel was required to offer peace (v. 10). Even when war became unavoidable, the Israelites were commanded not to destroy fruit trees (v. 19–20), symbolizing God's concern for life's ongoing provision. The military ethic of ancient Israel thus reflected God's sovereign control over violence—it was never self-willed or vengeful, but judicial and measured.

Old Testament scholar Walter C. Kaiser Jr. notes that these divine commands highlight not moral inconsistency but moral coherence within the covenant context: "God's sovereignty over life and death underscores His right to execute judgment through human agents when it serves His redemptive purposes."[1] In other words, when God authorizes judgment, He does so not as a cosmic tyrant but as the rightful ruler of creation, restoring moral order to a corrupted world.

From this perspective, killing becomes a theological issue, not merely a moral one. It is never humanity's prerogative to decide who lives or dies; that authority belongs to God alone. When He delegates it—whether through the state or through divinely appointed missions—it remains bounded by divine justice. Israel's wars were not fought for conquest or glory but as instruments of covenant preservation. They functioned as a means to protect the chosen people through whom the promise of redemption would one day arrive.

This framework helps modern believers understand that the Bible does not condone violence indiscriminately but situates it within the greater story of divine justice. God's acts of judgment—whether the flood in Genesis, the plagues of Egypt, or the conquest of Canaan—stemmed from His holiness, not from cruelty. Humanity's persistent rebellion required divine correction; otherwise, wickedness would perpetuate unchecked.

John Goldingay explains this tension well: "The paradox of God's justice and mercy lies not in contradiction but in covenantal faithfulness—

---

[1] Walter C. Kaiser Jr., *Toward Old Testament Ethics* (Grand Rapids, MI: Zondervan, 1983), 75.

He permits destruction only to preserve life."[2] God's wrath in Scripture always arises from love's demand for righteousness. His justice serves restoration, not annihilation. Even in judgment, He makes provision for mercy, as seen in His sparing of Rahab's family amid Jericho's destruction (Josh. 6:25).

The ethical question for the believer, therefore, is not whether killing has ever been permitted but under what circumstances it can align with divine intent. Scripture suggests two guiding principles. First, intent: killing becomes sin when motivated by hatred, greed, or revenge. Second, authority: legitimate use of lethal force requires proper divine or civil authorization. The Hebrew prophets constantly warned Israel that when its kings used violence for personal ambition, they invited judgment upon themselves (cf. 2 Sam. 12:9–10; Amos 1–2). Thus, God's allowance of killing is always conditional—anchored in justice, constrained by mercy, and executed through divine sovereignty.

When read in light of covenant theology, the Old Testament presents a divine warrior not as a brute force but as a righteous defender. God Himself is called "a man of war" (Exod. 15:3), yet His battles are aimed at establishing peace and righteousness. His wars are redemptive, not vindictive. Even His commands for Israel's conquest serve as acts of purification to protect the unfolding story of salvation that would culminate in Christ.

For modern military believers, this theological foundation provides both comfort and caution. Comfort, because it affirms that service under lawful authority can coexist with faith in a just God. Caution, because it reminds us that every use of force carries moral weight and must reflect God's justice, not our anger. The believing warrior stands in this tension —called to act with courage but governed by conscience.

Throughout history, soldiers of faith have wrestled with this same dilemma. The psalmist David, himself a warrior-king, often cried out for divine guidance before going into battle. In Psalm 144, he calls God his "trainer of hands for war," acknowledging dependence upon the Lord

---

[2] John Goldingay, Old Testament Theology, Volume 3: Israel's Life (Downers Grove, IL: InterVarsity Press, 2009), 234.

even for martial skill. Yet in Psalm 51, he confesses the moral gravity of bloodshed, pleading for a clean heart. David's life embodies the paradox of righteous warfare: courage rooted in humility, strength tempered by repentance.

Modern warfare has changed its technology but not its moral essence. The tools may be digital or nuclear, but the question remains ancient: can a believer take life without losing his soul? The Old Testament answers with realism and reverence: yes—if done under divine principle, with just cause, and in pursuit of peace rather than conquest.

For the believing warrior, this theological framework reaffirms that military service, rightly understood, is not rebellion against God's commandment but alignment with His justice. The Christian soldier does not glorify violence but disciplines it, turning strength into stewardship. As in Israel's story, the faithful in uniform serve not for personal glory but to protect life, restore order, and reflect the holy character of the One who alone holds the power over life and death.

"You Shall Not Murder": An Exegetical Focus

Few verses in Scripture have generated more moral debate than Exodus 20:13: "You shall not kill." In Hebrew, the command consists of only two words—לֹא תִרְצָח (lo' tirtsach)—yet these syllables carry centuries of ethical weight. For believers in the military, this command sits like a stone in the conscience: How can one wear the uniform of a nation authorized to use lethal force and still honor God's law that forbids killing? To answer that question faithfully, one must look beneath the surface of translation into the deeper current of Hebrew meaning, historical context, and theological intent.

The Hebrew Term "Ratsach" — A Narrow Prohibition

The word ratsach appears about forty-seven times in the Hebrew Bible, and its usage reveals a critical distinction. It consistently refers to murder—the unlawful, premeditated, or malicious taking of life—not to all killing in general. Ancient Israel, like every functioning society, recognized situations where taking life was permissible or even commanded under divine law: execution of murderers (Gen. 9:6; Num. 35:16-21), warfare authorized by God (Deut. 20), and self-defense (Exod. 22:2). Thus, the Sixth Commandment is not an absolute ban on killing; it is a moral boundary against unjust violence.

Modern readers often conflate "kill" with "murder," but in Hebrew, the distinction is decisive. Harag and nakah are broader verbs used for killing in war, accident, or even animal sacrifice. Ratsach, however, carries legal and ethical nuance—it is a moral offense, not merely a physical act. The intent behind the action determines its classification. Murder is motivated by hatred, vengeance, or gain; killing under divine or lawful authority may be an act of justice.

The ancient rabbis reinforced this distinction. In the Talmud (Sanhedrin 57a), they explain that ratsach condemns "blood guilt" between individuals, not judicial or wartime execution. The commandment was given to preserve life's sanctity, not to paralyze justice. God's law balances prohibition with protection—it guards both the victim and the righteous avenger.

Context within the Decalogue

The Sixth Commandment sits within the second table of the Decalogue, which governs human relationships. It follows prohibitions against idolatry, blasphemy, and dishonoring parents, implying that respect for life flows from reverence for God. To violate this command is to attack the image of God Himself (Gen. 9:6). Yet, the commandment's brevity—just two words—invites interpretation. Why such simplicity? Because it functions as a universal moral anchor: all human life is sacred, yet all justice belongs to God.

In Israel's covenantal society, this command was not isolated but integrated into a complex system of jurisprudence that included cities of refuge (Num. 35:9-34). These sanctuaries protected those who killed unintentionally, acknowledging both human fallibility and divine mercy.

Even in cases of justified killing, the law required rituals of cleansing (Deut. 21:1-9), symbolizing that taking life—even lawfully—remained a sobering act before God. The message was clear: human life may sometimes need to be taken, but it must never be treated lightly.

A Theological Reading of Exodus 20:13

To interpret lo' tirtsach properly, one must view it through God's moral nature. God is the giver and sustainer of life (Deut. 32:39). His command against murder reveals His desire for a society marked by justice, not vengeance. The prohibition thus upholds three theological truths:

1. *Life is sacred* because humanity bears God's image.
2. *Justice is divine*, not human; it must mirror God's character.
3. *Violence is restrained*, never glorified, within God's moral order.

God's command is not merely legal—it is relational. It calls His people to reflect His holiness in their dealings with others. Even when God authorizes Israel to go to war, He regulates conduct to preserve moral order. In Deuteronomy 20, soldiers are exempt if newly married, recently engaged, or afraid, demonstrating divine compassion even amid conflict.

As Walter C. Kaiser Jr. observes, "The purpose of the Sixth Commandment was not pacifism but protection. It stands as a safeguard for the sanctity of life against the chaos of human aggression."[3] This distinction means the believing warrior can honor this command even while bearing arms, provided that his actions stem from obedience, not hatred.

From Sinai to the Cross: Continuity and Fulfillment

Jesus did not abolish the commandment; He deepened it. In the

---

[3] Walter C. Kaiser Jr., *Toward Old Testament Ethics* (Grand Rapids, MI: Zondervan, 1983), 75.

Sermon on the Mount, He declares, "You have heard that it was said to those of old, 'You shall not murder,' ... but I say to you that everyone who is angry with his brother will be liable to judgment" (Matt. 5:21-22, ESV). Christ moves the discussion from behavior to motive. The moral battlefield shifts from the external act of killing to the internal act of hate.

By doing so, Jesus reclaims the original spirit of the commandment: the preservation of life through love. The believer is called not only to avoid physical murder but to root out the seeds of violence in the heart—bitterness, prejudice, and malice. Yet, this inner transformation does not negate the necessity of justice. When Jesus praised the faith of the Roman centurion (Matt. 8:5-13), He did not demand his resignation from military service. The centurion's faith coexisted with his duty.

Donald B. Kraybill captures this paradox in The Upside-Down Kingdom: "Jesus' radical ethic of love overturns vengeance, yet it does not abolish responsibility. Love confronts evil not by withdrawal, but by courageous engagement."[4]

For the believer in uniform, this means that following Christ entails both compassion and courage. The commandment forbids hatred, not defense; vengeance, not justice. To love one's enemy does not mean to allow the innocent to perish—it means to act without hatred, even when force is required.

Murder, Justice, and the Imago Dei

Every ethical reflection on killing must begin with the imago Dei—the truth that humanity reflects God's image. Genesis 9:6 grounds the prohibition of murder in this reality: "Whoever sheds human blood, by humans shall their blood be shed; for in the image of God has God made mankind." Here, divine justice is both protective and punitive. It defends the sanctity of life by delegating authority to human agents to restrain evil.

In this sense, the Sixth Commandment does not deny the legitimacy of civil justice; it affirms it. God entrusts governing authorities to maintain

---

[4] Donald B. Kraybill, *The Upside-Down Kingdom* (Scottdale, PA: Herald Press, 2003), 56.

order through measured coercion. This theology later surfaces in Paul's teaching in Romans 13:1-4, where the state "does not bear the sword in vain." The sword, an instrument of lethal force, becomes a symbol of lawful authority, not cruelty.

Yet, the believer's challenge remains: to distinguish between righteous authority and sinful aggression. Throughout Scripture, God's people are called to discern the difference between divinely sanctioned justice and self-serving violence. The prophet Micah summarizes divine expectation succinctly: "To do justice, to love mercy, and to walk humbly with your God" (Mic. 6:8). A soldier who carries this triad in his heart serves not only his nation but his Creator.

Exegetical Parallels in Ancient Law Codes

The Hebrew prohibition against murder did not emerge in isolation. Other Near-Eastern law codes, such as the Code of Hammurabi, also addressed homicide but grounded it in human retribution, not divine holiness. Israel's law, by contrast, locates morality in God's character, not social convenience. The Sixth Commandment reflects a theocentric ethic: life is sacred because it belongs to God, not because society deems it useful.

This theological foundation distinguishes biblical ethics from humanistic ethics. In human systems, morality evolves with circumstance; in divine law, it flows from God's immutable nature. Thus, when believers wrestle with moral dilemmas—like taking life in warfare—they are not guided by emotion or majority opinion but by revelation. Here lies the heart of biblical realism: God's law acknowledges human violence but sets boundaries to transform it into justice. The purpose of the commandment is not to create a pacifist nation but a holy one—a people who restrain their power under divine authority.

Practical Implications for Believers in Uniform

For Christians serving in the armed forces, this exegetical clarity brings both comfort and accountability. It affirms that lawful service does not violate God's law when performed under just authority and with righteous intent. Yet it also demands introspection. Every believer must examine motive and conscience before God. The uniform does not absolve moral responsibility—it magnifies it.

When a soldier acts under orders consistent with justice, he participates in God's restraining grace against evil. But when he acts out of hatred or vengeance, he crosses from divine service into personal sin. Thus, the believing warrior's obedience must always be filtered through the dual lenses of Scripture and conscience.

The battlefield may desensitize, but the Spirit humanizes. In moments of conflict, the Christian soldier remembers that even the enemy bears the imago Dei. The goal is not destruction for its own sake but protection of life, restoration of peace, and defense of the innocent. As Reinhold Niebuhr later argued, "The unwillingness to resist evil with force may be moral sentimentality rather than moral sensitivity."[5] The believing warrior must therefore walk the razor's edge between justice and compassion, courage and restraint.

For The Believing Warrior, the Sixth Commandment is not a barrier to military service but a blueprint for integrity within it. The Hebrew ratsach warns against unlawful killing, not against lawful defense. It calls believers in uniform to hold life sacred even as they bear the tools of war. The Christian soldier embodies paradox—armed yet merciful, disciplined yet compassionate, fierce in defense yet tender in spirit.

In honoring this commandment, the believer reflects the moral heart of God: a warrior who fights only to protect, a servant who carries the sword only to uphold peace. To "not murder" means to act always from love, even when duty demands strength. In this, the believing warrior stands not in contradiction to Scripture but in fulfillment of it—an agent of justice in a fallen world, guided by conscience, anchored in grace, and accountable to the Commander of Heaven's hosts.

---

[5] Reinhold Niebuhr, *Moral Man and Immoral Society: A Study in Ethics and Politics* (New York: Charles Scribner's Sons, 1932), 189.

The New Testament: A Call to Peace

The New Testament shifts the moral conversation from law to love, from external compliance to internal transformation. When Jesus of Nazareth ascended the hills of Galilee and opened His mouth to teach, He did more than interpret the Torah—He fulfilled it (Matt. 5:17). His words reoriented the moral compass of the faithful, pointing beyond the letter of the law toward its divine intention: reconciliation, mercy, and peace.

For the believing warrior, the New Testament represents both comfort and challenge. Comfort, because it reveals a Savior who understands the cost of violence; challenge, because it calls His followers to embody peace in a violent world. The Sermon on the Mount—perhaps the most quoted and misunderstood discourse in Christian ethics—stands as the focal point of this tension. It commands believers to turn the other cheek, love their enemies, and bless those who persecute them. But how are these ideals to be lived out by someone who has sworn to defend others, even at the cost of life?

Jesus and the Ethic of Peace

At the heart of Jesus' teaching lies the Kingdom of God—a new order breaking into the old. In this Kingdom, vengeance is replaced by forgiveness, and power is redefined as service. The beatitudes announce this reversal: "Blessed are the peacemakers, for they shall be called the children of God" (Matt. 5:9, KJV). The term eirēnopoioi ("peacemakers") describes those who actively cultivate reconciliation, not those who avoid conflict. Peace in Jesus' vocabulary is not passive absence of war but the active presence of justice.

Yet the peace of Christ is costly. It demands self-denial, humility, and moral courage. Jesus Himself lived this paradox—He came to bring peace, yet declared, "Do not think that I came to bring peace on earth; I did not come to bring peace but a sword" (Matt. 10:34, NKJV). This "sword" is not a literal call to arms but a metaphor for division—the inevitable tension that arises when divine truth confronts human rebellion. The Gospel of peace is, in itself, a kind of war against falsehood, hypocrisy, and sin.

In Jesus' confrontation with evil, we find a divine pattern for human action: restraint in motive, righteousness in purpose, and sacrifice in method. His refusal to retaliate at His arrest (Matt. 26:52-54) was not weakness but obedience; He surrendered His life to achieve redemption, not because resistance was immoral but because submission was necessary for salvation.

Paul's Theology of Government and Justice

If the Gospels reveal the heart of divine peace, Paul's epistles articulate its structure in human society. Romans 13:1–4 provides the New Testament's most explicit framework for understanding the moral legitimacy of governmental authority. "For there is no authority except from God, and those that exist have been instituted by God. … For he is the servant of God, an avenger who carries out God's wrath on the wrongdoer." The apostle does not glorify violence; he dignifies lawful authority as a tool of divine order.

This passage, often misunderstood, does not absolve rulers of accountability. It establishes that their authority is delegated, not autonomous. When governments use force to restrain evil, they serve God's justice; when they abuse power, they invite His judgment. For the Christian soldier, this distinction is critical. Obedience to authority remains a virtue—but not blind obedience. The believer must discern when orders align with justice and when they cross into sin.

Paul's realism recognizes the necessity of coercive power in a fallen world. Humanity's depravity demands systems of restraint. In that sense, the soldier's sword functions as an instrument of divine mercy, not vengeance—it protects the innocent from greater violence. This principle forms the foundation of the "just war" tradition developed by early theologians like Augustine and later refined by Aquinas. Both argued that warfare, though tragic, could be morally justified when it meets certain conditions: legitimate authority, just cause, right intention, and proportional means.

Reinhold Niebuhr revisited this tension in the twentieth century, describing it as Christian realism. He wrote, "The unwillingness to use force to restrain evil in the name of love can become a betrayal of love itself."[6]

Niebuhr's realism does not celebrate violence—it acknowledges it as a necessary tool in a world marred by sin. Peace without justice, he argues, is not peace but surrender to evil.

The Apostolic Example of Service and Sacrifice

The New Testament contains no direct condemnation of soldiers or military service. On the contrary, soldiers appear in the Gospel narrative as individuals capable of faith, integrity, and conversion. The Roman centurion in Capernaum (Matt. 8:5–13) is commended by Jesus for his faith, not rebuked for his profession. Another centurion, Cornelius, becomes the first Gentile convert in Acts 10—a man described as "devout and God-fearing." Neither is told to abandon his duty.

These examples suggest that the moral problem is not the profession of arms but the heart behind the action. The New Testament recognizes that even within a corrupt system, individuals can serve righteously. Soldiers in the first century occupied a complex social space—agents of Roman authority, yet often despised by the oppressed Jewish populace. Still, Scripture presents them as capable of faith, generosity, and obedience to God.

When John the Baptist preached repentance to the crowds, soldiers asked him, "What should we do?" His reply was straightforward: "Do not extort money and do not accuse people falsely—be content with your pay" (Luke 3:14, NIV). Notice what John did not say: he did not tell them to leave the military. Instead, he called them to moral integrity within it.

This instruction captures the essence of Christian vocation in the

---

[6] Reinhold Niebuhr, *Moral Man and Immoral Society: A Study in Ethics and Politics* (New York: Charles Scribner's Sons, 1932), 189.

public sphere: not withdrawal, but transformation. A believer's call is not to escape the world but to bear witness within it. Whether priest or pilot, chaplain or infantryman, each disciple serves the same Master and must carry the same cross-shaped ethic into their vocation.

The Paradox of Peace Through Sacrifice

The New Testament's portrayal of peace reaches its climax in the crucifixion. The cross stands as the definitive act of divine nonviolence and the most violent event in history. God absorbs humanity's aggression without retaliation, transforming death into redemption. Herein lies the paradox: peace is achieved not by avoidance of conflict but by confronting evil with sacrificial love.

This is why Christian theology can affirm the legitimacy of protective force while condemning vengeance. The goal is not destruction but restoration. The believer who defends others participates—imperfectly but meaningfully—in God's redemptive justice. Every act of defense rooted in love echoes the divine initiative to protect creation from chaos.

In this sense, peace is not fragile tranquility but redeemed order. The Greek word eirēnē (peace) derives from eiro, "to join or bind together." Peace, then, is relational wholeness—the reuniting of what sin has torn apart. The Christian soldier, when guided by righteousness, becomes an agent of this restoration, binding together nations, protecting lives, and preserving the moral fabric of society.

Jesus and the Sword: A Balanced Understanding

The Gospel accounts record two seemingly contradictory moments involving swords. In Luke 22:36, Jesus instructs His disciples, "Let the one who has no sword sell his cloak and buy one." Yet moments later, when Peter uses that sword to defend Him, Jesus commands, "Put your sword back into its place. For all who take the sword will perish by the sword" (Matt. 26:52, ESV).

How can both statements be true? The answer lies in context. Jesus' command to arm themselves was symbolic—a call to spiritual readiness amid coming persecution. The sword represents vigilance, not aggression. Peter's mistake was not carrying the sword but misusing it. He fought a spiritual battle with carnal means, attempting to prevent a crucifixion that would bring salvation.

The principle for the believer is timeless: possession is not permission. Having the capacity for violence does not justify its use apart from divine purpose. The Christian soldier must, therefore, master both restraint and readiness. His weapon is not an extension of rage but of responsibility—a tool of justice wielded under authority, not emotion.

The Role of Conscience and Calling

The Apostle Paul writes, "Let each one remain in the calling in which he was called" (1 Cor. 7:20, NKJV). For some believers, that calling may indeed be to military service. The conscience becomes the moral compass for discernment. Paul's framework in Romans 14 affirms that believers may hold differing convictions regarding disputable matters, provided each acts "unto the Lord."

For the pacifist, faithfulness may mean refusing to bear arms; for the warrior, it may mean bearing them responsibly. Both must act from conviction, not convenience. The danger lies not in disagreement but in judgment—condemning another servant's obedience because it differs from one's own.

This balance honors the diversity of callings within the Body of Christ. Not all are called to fight, but all are called to love. The believer in uniform must see his duty not as contradiction but as commission—to protect the weak, uphold justice, and model virtue amid chaos.

Christian Peace in a Violent World

Miroslav Volf in Exclusion and Embrace reminds us that Christian peace is not achieved by isolation but through radical inclusion: "Peace requires the embrace of the other—not withdrawal from the world, but reconciliation within it."[7] This theology of embrace transforms how we understand conflict. To "make peace" does not mean to tolerate injustice but to restore broken relationships through love, even when it costs us dearly.

In a military context, this peace may take the form of discipline under fire, mercy toward enemies, and integrity under pressure. Every act of restraint, every decision to spare rather than destroy, every humanitarian mission carried out in war-torn regions bears witness to Christ's reconciling love. The believing warrior becomes a paradoxical presence: one who carries instruments of war but embodies the spirit of peace.

This is the essence of Christian service—to fight not for dominance but for dignity, not for conquest but for compassion. When violence becomes necessary, it is mourned, not celebrated. The soldier weeps for what he must do but finds comfort in knowing he fights to preserve the peace others cannot defend for themselves.

For The Believing Warrior, the New Testament's call to peace is not a rejection of service but a redemption of it. Christ's command to love one's enemies becomes the heartbeat of the Christian in uniform, transforming every mission into a ministry of protection and peacekeeping. The believing warrior stands not as a contradiction to the Gospel but as its embodiment in the harshest arenas of life—a living testimony that divine peace is not the absence of conflict but the presence of righteousness amid it.

The cross reminds us that the cost of peace is sacrifice. And the soldier who serves with faith, humility, and moral clarity reflects the very nature of the One who gave His life to reconcile heaven and earth. To kill is never to be desired—but to defend, to protect, and to preserve life can be an act of worship when done under God's authority and love.

God's Ethics vs. Human Ethics

If there is a single distinction that defines divine morality from human

---

[7] Miroslav Volf, *Exclusion and Embrace: A Theological Exploration of Identity, Otherness, and Reconciliation* (Nashville, TN: Abingdon Press, 1996), 117.

morality, it is this: God's ethics flow from perfect love and omniscient justice, while human ethics emerge from partial understanding and fallen nature. The difference is not simply in degree—it is categorical. Human ethics are finite, reactive, and self-interested. God's ethics are infinite, proactive, and grounded in eternal wisdom. For the believing warrior, understanding this divide is not philosophical luxury—it is moral necessity. Every trigger pulled, every order obeyed, every life spared or lost must ultimately be filtered through the difference between divine will and human impulse.

Divine Justice: A Reflection of God's Nature

The starting point for understanding divine ethics lies in God's own character. Scripture affirms repeatedly that "all His ways are justice; a God of faithfulness and without iniquity, just and upright is He" (Deut. 32:4, ESV). God's justice is not a subset of His being—it is an extension of His holiness. Everything He commands proceeds from perfect moral purity, unmarred by the selfish ambitions that stain human judgment.

This perfection is what allows God to execute judgment without contradiction. When He judges, He does not merely balance scales; He restores harmony to a disrupted creation. His justice is restorative, not merely retributive. Theologian John Goldingay highlights this divine balance: "God's justice functions as the guardian of life. He is not the destroyer but the restorer of moral order."[8] In other words, divine justice always serves the preservation of creation, even when it involves destruction of wickedness.

Humans, by contrast, are bound by perspective. Our justice—no matter how noble—is limited by bias, emotion, and self-preservation. We see through "a glass darkly," as Paul wrote (1 Cor. 13:12). We judge based on fragments of information, influenced by fear or pride. This is why Scripture warns so emphatically against vengeance. "Do not take revenge, my dear friends, but leave room for God's wrath," Paul insists, "for it is

---

[8] John Goldingay, Old Testament Theology, Volume 3: Israel's Life (Downers Grove, IL: InterVarsity Press, 2009), 234.

written: 'Vengeance is mine, I will repay,' says the Lord" (Rom. 12:19, NIV). Vengeance corrupts because it attempts to seize what belongs only to God—His prerogative to determine right and wrong without error.

The Old Testament's ethical structure reflects this truth. Every law regarding warfare, punishment, or purification presupposes that God alone is the moral authority. Israel was forbidden from taking up arms unless explicitly authorized by divine decree (Deut. 20:4). This principle ensured that even in the context of violence, the moral compass pointed heavenward. The moment killing became an act of personal retribution, it ceased to be divine justice and became human sin.

Human Ethics: The Weight of Fallenness

Human ethics, on the other hand, are the tragic echo of Eden. Ever since Adam and Eve ate from the Tree of Knowledge of Good and Evil, humanity has claimed the right to define morality independently from God. That act of rebellion birthed moral relativism—the belief that good and evil can be determined apart from divine revelation. The serpent's whisper, "You will be like God," (Gen. 3:5) still resonates in modern conscience and culture alike.

This self-referential ethic has led to both moral confusion and contradiction. Humans long for justice but resist the authority of the One who defines it. Our wars often begin in pride and end in pain because we fight for dominance rather than righteousness. Yet even in our moral frailty, the image of God endures, urging humanity toward justice, mercy, and truth.

For the military believer, this tension is acute. The call to obey lawful orders must always be weighed against allegiance to divine law. Blind nationalism cannot replace moral discernment. Soldiers throughout history have justified atrocities under the banner of obedience, forgetting that legality is not synonymous with righteousness. To act ethically within the military framework, a Christian must discern not only what is permitted but what is holy.

This is precisely why Miroslav Volf argues that Christian ethics must

BE "cruciform"—shaped by the cross rather than by convenience.[9] Christ's ethic of self-giving love exposes the limits of human justice. Whereas humanity demands repayment, God offers redemption. Where humans seek control, God demonstrates surrender.

The Problem of Moral Distance in Warfare

Modern warfare compounds the moral challenge by introducing distance between the soldier and the act. In ancient times, combat was immediate—swords clashed, eyes met, and every life taken was personal. Today, technology has abstracted killing into data and coordinates. A missile launched from thousands of miles away can erase lives unseen. This detachment risks dulling the conscience, turning moral weight into mechanical procedure.

For the believing warrior, moral vigilance must remain acute even in technological detachment. The further we are removed from the physical presence of those affected by our actions, the greater our responsibility to maintain spiritual awareness. The Christian soldier must remember that even when justice requires lethal force, every life lost carries sacred value before God. Moral numbness is not neutrality—it is decay.

Herein lies the tension between divine and human ethics once again. God never loses sight of the person behind the sin. His justice is always relational; it seeks to heal what is broken, not merely punish wrongdoing. Human systems, however, often depersonalize evil, reducing the enemy to a category rather than a creation. The Christian must resist this impulse. Every adversary remains a soul for whom Christ died. The true enemy is never flesh and blood but the spiritual forces of darkness that corrupt and destroy (Eph. 6:12).

Reconciling Divine and Human Ethics in Service

Bridging the gap between divine justice and human responsibility

---

[9] Miroslav Volf, *Exclusion and Embrace: A Theological Exploration of Identity, Otherness, and Reconciliation* (Nashville, TN: Abingdon Press, 1996), 117.

requires humility and prayer. The believing warrior stands as a moral ambassador, called to reflect heavenly values within earthly systems. The military profession—structured, hierarchical, and mission-driven—can easily drift into moral absolutism, where orders replace conscience. But Scripture calls believers to a higher allegiance.

When Peter and the apostles were commanded to remain silent about Jesus, they responded, "We must obey God rather than men" (Acts 5:29). This was not rebellion; it was fidelity. In the same spirit, the Christian service member must uphold integrity even when it risks reputation or rank. Ethical obedience is not blind compliance—it is faithful discernment. The military code of honor, when rightly understood, can harmonize beautifully with divine principles: courage, loyalty, respect, integrity, and selfless service are virtues rooted in biblical truth.

Reinhold Niebuhr frames this paradox through the lens of "moral man within immoral society." He explains that individual virtue cannot fully redeem collective sin but must still strive toward it.[10] The believer's task, therefore, is not to perfect the system but to infuse it with light. The Christian soldier cannot cleanse war of its horror, but he can conduct himself honorably within it—bearing witness to the God who redeems even humanity's darkest moments.

This kind of moral realism guards against despair. It recognizes that while no human system will achieve God's perfect justice, divine grace can still work through flawed instruments. The soldier's service becomes sanctified not because the battlefield is holy, but because the believer's heart is surrendered to the Holy One who sends him there.

When Human Ethics Fail: A Need for Redemption

History offers tragic reminders of what happens when human ethics operate without divine restraint. From the crusades to totalitarian regimes, countless atrocities have been committed in the name of righteousness detached from revelation. Humanity's moral compass, untethered from

---

[10] Reinhold Niebuhr, *Moral Man and Immoral Society: A Study in Ethics and Politics* (New York: Charles Scribner's Sons, 1932), 189.

God, inevitably points inward rather than upward.

The solution is not to abandon the pursuit of justice but to re-anchor it in divine truth. God's ethics cannot be improved upon because they arise from omniscience. He knows not only the action but the motive, not only the sin but the wound behind it. Where human courts can only punish, divine justice can restore.

For the believer in uniform, this means viewing every decision—whether in combat or command—through the lens of compassion. Justice and mercy are not enemies; they are twin pillars of divine ethics. The prophet Micah captures this tension perfectly: *"He has shown you, O man, what is good; and what does the Lord require of you but to do justly, love mercy, and walk humbly with your God?"* (Mic. 6:8, NKJV).

True justice, then, must always be tempered by mercy, for without it, righteousness becomes tyranny. And mercy must never erase justice, or compassion becomes complicity. The believing warrior must hold both with trembling hands, knowing that he serves a God who does the same.

Ethics as Worship

Ultimately, ethics for the Christian is not merely about decision-making—it is about worship. Every moral choice either magnifies or diminishes God's image within us. When a soldier refuses to dehumanize his enemy, shows restraint under pressure, or protects the innocent at personal risk, he is performing an act of worship. His obedience in the field becomes as sacred as a hymn sung in a sanctuary.

This transforms military ethics from mere compliance to consecration. The battlefield, with all its peril, becomes an altar where faith is tested and revealed. It is where the soldier's theology becomes embodied action, where words about love and justice must take flesh under fire. When a believing warrior acts justly and mercifully amid chaos, he preaches a silent sermon that echoes louder than any words could express.

The Divine Perspective on Violence

Scripture reminds us that God's final goal is not destruction but redemption. The entire biblical narrative moves from violence to peace—from the blood of Abel crying out from the ground (Gen. 4:10) to the redeemed multitudes singing, "Worthy is the Lamb who was slain" (Rev. 5:12). Divine justice begins in wrath but ends in restoration.

The cross stands as the ultimate convergence of these two ethics. There, divine judgment and divine mercy meet. Human violence pierces the body of Christ, but God transforms that violence into salvation. The very instrument of death becomes the means of life. In this cosmic reversal, believers find their ethical pattern: to confront evil not with hatred but with holy courage, to participate in justice without forfeiting grace.

For the Christian in uniform, this means carrying the cross into conflict—living as one who understands both the weight of sin and the hope of redemption. Every mission, every duty, every oath must be infused with the awareness that divine ethics are not about domination but restoration.

For The Believing Warrior, understanding the difference between God's ethics and human ethics is crucial to serving with both strength and sanctity. Divine ethics remind the soldier that the ultimate goal of his duty is not destruction but defense—not conquest but compassion. Human warfare may take life, but divine purpose can give it meaning.

The believer who serves under authority yet bows to a higher King reflects the paradox of Christ Himself—obedient unto death, yet conqueror through love. The Christian warrior's task is not to erase conflict but to redeem it, to stand as a moral bridge between heaven's justice and earth's brokenness. In doing so, he transforms his uniform into a symbol of divine mercy, proving that even in war, grace can still wear boots.

The Opportunity to Give Life

To speak of killing without speaking of life would be an incomplete theology. For the believing warrior, the battlefield is not merely a place where death occurs—it is also a place where life is preserved, defended, and sometimes miraculously restored. The act of serving in the military is often framed by outsiders as an occupation of violence, but those who have worn the uniform understand that it is, at its deepest level, a vocation of protection. To serve in the military is not primarily to take life; it is to safeguard it.

This truth becomes evident when one looks beyond the visible mechanics of war to the invisible motives that guide the heart of the Christian soldier. The believer in uniform does not fight because he loves conflict but because he loves peace. He does not defend because he despises the enemy but because he cherishes the innocent. In a world plagued by chaos, oppression, and evil, service becomes not an act of aggression but an act of stewardship—guarding the sanctity of life against forces that seek to destroy it.

The Call to Preserve Life

The notion that life can be given, not only through healing but through defense, finds deep roots in Scripture. In the Old Testament, defenders of Israel were viewed as instruments of divine preservation. King David's soldiers, known as the "mighty men," were not remembered for their bloodlust but for their courage in protecting the covenant people from destruction. Their valor secured the survival of a nation through which God's redemptive plan would unfold.

This protective instinct is not antithetical to faith—it is an extension of divine compassion. God Himself is described as a "shield" (Ps. 3:3), a "fortress" (Ps. 18:2), and a "strong tower" (Prov. 18:10). When believers assume similar roles on earth—protecting the weak, upholding justice, and restraining evil—they reflect the very attributes of their Creator. The military profession, when exercised under righteousness, becomes a living arable of God's protection.

In this sense, the Christian soldier mirrors the Good Shepherd, who "lays down his life for the sheep" (John 10:11, ESV). To stand between danger and the defenseless is an act of love, not of hatred. It is to imitate Christ's sacrificial posture in a fallen world, where peace is often purchased at great cost. As Donald B. Kraybill reminds us, "The upside-down kingdom of God measures greatness not by power or preservation of self, but by the willingness to serve and sacrifice for others."[11]

Desmond Doss: A Testament of Courage and Conviction

One of the most profound examples of giving life amid war is found in the story of Desmond T. Doss, the Seventh-day Adventist medic immortalized in The Unlikeliest Hero. Doss entered World War II with a Bible in one hand and unwavering conviction in his heart. As a conscientious objector, he refused to carry a weapon, choosing instead to heal rather than harm.

During the harrowing Battle of Okinawa, Doss faced unrelenting fire from enemy forces. While others sought cover, he remained exposed, determined to rescue the wounded. One by one, he carried men to safety—Americans and even some Japanese soldiers—praying after each rescue, "Lord, please help me get one more." By the end of the battle, he had saved seventy-five lives without firing a single shot.

His story transcends the boundaries of denomination or creed. It embodies what it means to serve God in a place of death while choosing to be an agent of life. Doss was awarded the Congressional Medal of Honor, yet his legacy is not of heroism alone but of holiness—faith lived out in the fiercest crucible of war. As Booton Herndon beautifully recorded, "He sought no glory, no vengeance, no conquest. His only weapon was faith; his only mission was mercy."[12]

Doss's witness reframes the conversation around Christian service. He

---

[11] Donald B. Kraybill, *The Upside-Down Kingdom* (Scottdale, PA: Herald Press, 2003), 56.

[12] Booton Herndon, *The Unlikeliest Hero* (Mountain View, CA: Pacific Press Publishing, 1967), 78.

reminds us that one's military vocation need not be defined by taking life but by valuing it. Even on the battlefield, a believer can reflect the compassion of Christ, proving that faith and service are not contradictions but complements when rooted in love.

Giving Life Through Acts of Service

Not every soldier is a medic, but every believer in uniform carries the potential to give life. Whether it is through humanitarian missions, disaster relief, or the quiet act of moral courage, service provides countless opportunities for restoration.

Consider the work of military engineers who rebuild war-torn villages, or chaplains who counsel the grieving and restore hope to the weary. Consider the pilots who deliver food to besieged regions, or the sailors who evacuate civilians from danger zones. These acts of mercy, though often overshadowed by combat, are among the truest expressions of what it means to serve in the image of God.

Each of these efforts reveals a vital truth: the mission of a believing warrior is not confined to victory but extends to redemption. The soldier's hands, though trained for battle, can also heal, build, and bless. The same discipline that enables them to destroy threats can also empower them to sustain life. The dichotomy between killing and saving, destruction and defense, finds resolution in the soldier who sees his duty as ministry—one who understands that the greatest victory is not in conquest, but in compassion.

The Power of Restraint

One of the most underestimated ways of giving life in warfare is through restraint. True strength is not measured by one's ability to strike but by the wisdom to hold back. In Scripture, David exemplified this when he refused to kill King Saul, despite having the opportunity. "The Lord forbid that I should do this thing to my master, the Lord's anointed," he declared (1 Sam. 24:6, NKJV). David's mercy toward his persecutor became an act of life-giving grace—a refusal to let vengeance overtake virtue.

In modern military ethics, restraint remains the mark of moral maturity. Rules of engagement, proportionality, and non-combatant immunity are not merely political constructs—they are theological principles rooted in the image of God. Every decision to spare rather than destroy is a whisper of divine mercy amid the clamor of war.

The believing warrior, guided by the Spirit, must recognize that restraint is not weakness but worship. Each act of mercy honors the Creator who values every soul. The ability to stop one's hand when vengeance tempts it is proof that divine ethics have triumphed over human impulse. In that moment, the soldier becomes a vessel of grace—a living contradiction to the chaos of war.

Life Beyond the Battlefield

For many veterans, the struggle to give life continues long after the last battle ends. The transition from combat to civilian life is often fraught with moral injury, trauma, and loss of identity. Yet even here, God's redemptive work continues. The same courage that carried them through war can now be used to bring healing, mentorship, and hope to others.

Miroslav Volf observes that reconciliation requires both the acknowledgment of pain and the embrace of the other.[13] Veterans who channel their experiences into service—whether through counseling, community work, or advocacy—become living agents of reconciliation. They transform suffering into empathy, guilt into grace, and memory into ministry. In doing so, they continue to "give life," proving that service to God and country does not end with discharge but endures through compassion.

The believer's witness after war may be even more powerful than his witness within it. When others see that one can endure horror yet retain faith, can experience loss yet still love, they encounter the very essence of

---

[13] Miroslav Volf, *Exclusion and Embrace: A Theological Exploration of Identity, Otherness, and Reconciliation* (Nashville, TN: Abingdon Press, 1996), 117.

resurrection—the victory of life over death.

Peace as the Ultimate Gift of Life

The ultimate act of giving life is not in sparing it temporarily but in leading others toward eternal peace. This is the soldier's deepest calling—to create spaces where families can flourish, children can dream, and faith can thrive. Peace is not a political achievement; it is a spiritual inheritance. It is the byproduct of justice secured and mercy extended.

The Christian soldier serves not merely to prevent war but to establish the conditions in which peace becomes possible. His sacrifice mirrors the work of Christ, who purchased our peace with His blood. The soldier's wounds become testimonies of love's cost; his scars, sacraments of service.

As the prophet Isaiah foretold, "The work of righteousness will be peace, and the effect of righteousness, quietness and assurance forever" (Isa. 32:17, NKJV). Peace is not the absence of battle—it is the fruit of righteous labor. The believing warrior participates in this divine vocation whenever he stands for truth, defends the defenseless, and restrains evil for the sake of life.

For The Believing Warrior, the call to arms is ultimately a call to life. The question "to kill or not to kill" finds its final resolution not in legality but in love. The believer who serves in uniform must remember that his weapon, his training, and his authority are tools of stewardship, not domination. His greatest victory is not in the number of battles won, but in the number of lives preserved.

The Christian soldier gives life in countless ways—by showing mercy when others expect malice, by rescuing rather than retaliating, and by bearing witness to a God who redeems even in the darkest hour. The cross itself was a battlefield, and there, amid suffering and death, life triumphed. The believing warrior follows that same pattern, proving that even in the theater of war, God's love can still wage peace.

When the guns fall silent and the flags are folded, what remains is not the memory of destruction, but the legacy of service—a life poured out so that others might live. That is the ultimate act of worship for the warrior who believes.

PART II

THE ELITE FORCES IN THE BIBLE

CHAPTER 6

Are Armed Forces Biblical?

ONE OF THE MOST COMMON objections among Christians regarding military service is the belief that war is inherently sinful—a symptom of a fallen, rebellious world. While it is true that sin has corrupted human institutions and the way warfare is often carried out, it is not entirely accurate to say that war itself originated with man. The Bible paints a broader, more profound picture. The first war ever recorded did not occur on earth but in heaven, long before humanity existed.

The First War Was Not on Earth

The Book of Revelation gives us a glimpse into this cosmic battle: "Then war broke out in heaven. Michael and his angels fought against the dragon, and the dragon and his angels fought back. But he was not strong enough, and they lost their place in heaven. The great dragon was hurled down—that ancient serpent called the devil, or Satan, who leads the whole world astray" (Revelation 12:7–9, NIV).

This passage is foundational. It introduces the reality that conflict—armed conflict—originated in the spiritual realm, not in human politics. The word "war" here comes from the Greek *polemos*, a term used elsewhere in scripture to describe real, organized battle. This is not symbolic. This is war involving angels, ranks, and leadership—led by Michael, the Archangel. According to Daniel 10:13 and 12:1, Michael is referred to as a "great prince" who defends God's people. Some theological traditions equate Michael with Christ in His pre-incarnate form—Commander of Heaven's Army.

The presence of war in heaven before human sin confronts us with a challenging truth: conflict is not inherently evil. It becomes evil when it is motivated by pride, malice, greed, or revenge—just as Satan's rebellion was. However, war in the defense of righteousness, led by God's appointed forces, reflects divine justice and the defense of what is good and holy.

This heavenly war was not caused by sin on earth but by rebellion in heaven. Satan, desiring equality with God (Isaiah 14:12–15), organized an uprising. It was not God who initiated this war, but He permitted His angels to respond with force—to defend truth, justice, and His kingdom's order. That moment became the template for every war of justice that followed.

From this starting point, we learn a powerful hermeneutical principle: not all warfare is unrighteous. If angels in heaven can fight for the cause of divine justice, then the question isn't whether war is biblical, but whether our participation is aligned with righteousness.

This cosmic reality reframes our understanding of earthly service. When a Christian soldier stands against evil—not in hatred, but in defense of the innocent—he is walking in the pattern of Michael and the angelic hosts. He is not acting outside of God's will, but very possibly within it.

In the next pages, we'll explore how this theme continues throughout the Old Testament and New Testament, ultimately culminating in the eschatological vision of final justice.

God, the Divine Warrior

Having seen that war originates in the heavenly realms, not merely in the depravity of fallen humanity, we must now consider how God Himself is portrayed as a warrior throughout Scripture. The Old Testament is filled with language and imagery of divine warfare—not merely metaphorical, but active intervention in human battles.

Yahweh Sabaoth – "The Lord of Hosts"

One of the most frequent and revealing titles for God in the Hebrew Bible is Yahweh Sabaoth, which translates as "The LORD of Hosts" or "The LORD of Armies." This name appears over 270 times, and it speaks directly to God's identity as the supreme military commander—over both heavenly beings and earthly forces. "The LORD is a warrior; the LORD is his name" (Exodus 15:3, CSB).

This declaration follows the deliverance of Israel from Pharaoh's army through the Red Sea. God didn't just orchestrate the escape; He orchestrated a military defeat. The text in Exodus 14 is explicit: "The LORD will fight for you." Here, we witness the divine act of warfare—not passive deliverance, but intentional destruction of an oppressive army for the sake of Israel's freedom.

Holy War and Divine Justice

The concept of "holy war" in the Old Testament is often misunderstood. Critics accuse God of endorsing violence, but such battles were never about conquest for conquest's sake. Rather, they were divine judgments on deeply corrupt and violent societies (see Genesis 15:16; Deuteronomy 9:4-5).

The military campaigns led by Joshua, Gideon, and David were never launched at their own discretion; they were guided, sanctioned, and often commanded directly by God. "When you go out to war against your enemies, and the LORD your God gives them into your hand…"

(Deuteronomy 20:1).

Notice: "When," not if. The assumption is that war would be part of Israel's experience—and that God Himself would be involved in the outcome. What matters is not merely the presence of war but the righteousness of the cause and the divine guidance behind it.

The Commander of the Lord's Army

One of the most striking moments in the Old Testament occurs in Joshua 5:13–15, when Joshua encounters a mysterious warrior with a drawn sword: "Joshua went up to him and asked, 'Are you for us or for our enemies?' 'Neither,' he replied, 'but as commander of the army of the LORD I have now come.'"

This divine figure, who many scholars identify as the pre-incarnate Christ, does not take sides in human politics. He represents God's higher purposes, which transcend human agendas. The significance here is staggering: God has an army, and that army has a Commander, who engages directly in the military struggles of His people.

Joshua's response is telling: he falls facedown in reverence, recognizing the holiness of this encounter. He is told to remove his sandals—echoing the call of Moses at the burning bush—further reinforcing that this is no ordinary warrior, but God Himself in a militant role.

Military Imagery and Theology

Psalm 24:8 asks: "Who is this King of glory? The LORD, strong and mighty, The LORD, mighty in battle." God is not described here in priestly robes or teaching in a synagogue—He is shown as a conquering king, sword in hand, entering His city in triumph. To reject the idea of righteous warfare is to reject a central image of who God is: a defender of the oppressed, a destroyer of evil, and a protector of His covenant people.

Jesus—Prince of Peace and Commander of Heaven's Armies

When people think of Jesus, they often envision the gentle teacher, the healer, the Lamb of God—images that emphasize meekness, mercy, and sacrifice. These are not wrong. But they are incomplete. The New Testament reveals another side of Jesus—one that is often overlooked, especially in theological discussions about violence and military service. Jesus is not only the Lamb; He is also the Lion. He is the Prince of Peace, but also the Commander of Heaven's armies.

Jesus in Revelation: Warrior King

Nowhere is this militant image more vivid than in Revelation 19:11–16: "Then I saw heaven opened, and behold, a white horse! The one sitting on it is called Faithful and True, and in righteousness he judges and makes war. His eyes are like a flame of fire…He is clothed in a robe dipped in blood, and the name by which he is called is The Word of God… From his mouth comes a sharp sword with which to strike down the nations."

This is not poetic symbolism alone. This is Jesus Christ, returning to Earth not to teach or heal, but to confront and destroy the armies of evil. The sword proceeding from His mouth represents the power of His word—the same word that spoke the world into existence now brings final judgment.

Here we find the paradox: Jesus wages war in righteousness. The Greek word used is polemeō, meaning to do battle. He is not driven by vengeance or hatred, but by a holy mission to eliminate evil. His robe dipped in blood is not a metaphor for His own sacrifice—He wears it before the battle begins. This is the blood of His enemies—a symbol of coming judgment.

This revelation transforms our understanding of Christ's mission. Peace is not simply the absence of conflict. Biblical peace (shalom) is the presence of justice, righteousness, and order. And sometimes, establishing that peace requires confronting darkness with divine force.

The Lamb Who Fights

Even during His earthly ministry, Jesus was not a pacifist in the modern political sense. He spoke of division (Luke 12:51), conflict (Matthew 10:34), and spiritual warfare (Luke 11:21–22). His cleansing of the temple (John 2:13–17) was a violent act—driving out the corrupt money changers with a whip. Was He lacking love? No. He was zealous for righteousness. His mission demanded confrontation, not just compassion.

In Matthew 24, Jesus speaks of wars, rumors of wars, and global unrest—not to condemn such things outright, but to explain that they are part of the unfolding of God's plan. And in Luke 22:36, when He tells His disciples to "sell your cloak and buy a sword," it is clear that Jesus acknowledges the need for protection and preparedness in a hostile world. He does not abolish the idea of defense—He reframes it within the context of kingdom purpose.

Spiritual and Physical Warfare

Jesus' teachings clearly elevate spiritual warfare as the primary battlefront. "My kingdom is not of this world," He told Pilate (John 18:36). But He did not say His kingdom was disconnected from this world. His strategy was higher—not to seize political power through force, but to win hearts through truth. Still, this did not negate the fact that war, even physical war, has a role in redemptive history.

In the same way, a Christian may serve in the military without contradicting the spirit of Christ's teachings. It depends not on the presence of violence, but on the motive, purpose, and heart behind it.

Just as Jesus came with a mission of love and also with a sword of judgment, so too believers can enter the battlefield not as agents of destruction, but as warriors of righteousness.

Eschatology and the Final Eradication of Evil — The War to End All Wars

If the Bible begins with a war in heaven, it ends with one too. Between Genesis and Revelation lies not only the story of humanity but the story of God's military campaign to eradicate evil forever. While human wars are often driven by pride, greed, or vengeance, the divine war is a righteous confrontation against sin, rebellion, and the destruction of life itself. The culmination of this war, found in the eschatological visions of Revelation, brings clarity to the question: Is war always wrong, or can it be holy?

The Thousand Years and the Final Judgment

Revelation 20 outlines the chronology of the final phases of God's military strategy:

1. Satan is bound for 1,000 years (Revelation 20:1–3).
2. The saints reign with Christ (Revelation 20:4–6).
3. Satan is released and gathers armies (Revelation 20:7–9).
4. Fire from heaven destroys them all.
5. The Great White Throne Judgment takes place (Revelation 20:11–15).

This is no metaphor. It is not a battle of opinions or ideologies. It is a cosmic military campaign. Satan does not go down quietly—he rallies the nations, symbolized as Gog and Magog, for one last war. He is the archenemy, the commander of the forces of darkness. His followers, both spiritual and human, align themselves for battle.

And yet, no sword is swung by the saints. No counterattack is launched. Why? Because the victory belongs to God alone. "But fire came down from heaven and consumed them" (Revelation 20:9). The war is over in a flash of holy justice.

This scene is the climax of spiritual and physical warfare in the Bible.

It confirms a sobering truth: war has a place in God's redemptive strategy, but only as a temporary means to a final and everlasting peace.

The Final Eradication of Sin

The purpose of this ultimate war is not conquest—it is cleansing. God's war is never about power for its own sake. It is about truth, justice, and the restoration of creation. After this final judgment, sin will never rise again. As Nahum 1:9 promises, "What do you conspire against the Lord? He will make an utter end of it. Affliction will not rise up a second time."

This is crucial for understanding why military service can reflect a divine calling. The Bible does not glorify violence, but it also does not shy away from using warfare as a tool of righteous judgment. In the end, God does not send politicians or philosophers to settle the final conflict—He leads armies. And it is through this war that evil is forever destroyed.

Heaven's Veterans: The Redeemed Warriors

Interestingly, Revelation does not only depict angels and divine beings as warriors. The saints themselves are referred to as "those who overcome" (Revelation 2–3). The Greek word nikaō means "to conquer" or "to prevail in battle." These aren't passive spectators—they are spiritual veterans, tested by fire and faithful in war.

"And they overcame him by the blood of the Lamb, and by the word of their testimony; and they loved not their lives unto the death" (Revelation 12:11).

This is not just poetic language. It's a war cry. The saints are spiritual soldiers, marked not by medals but by faithfulness in the face of spiritual war.

God's army is made up of those who fought the good fight of faith, whether with literal weapons or spiritual armor. In the end, they will stand with Christ, not because they were pacifists, but because they were courageous, disciplined, obedient, and faithful.

Why Military Service Can Reflect God's Nature

So, is it biblical to serve in the armed forces?

After examining the full scope of scripture—from the celestial war in heaven to the apocalyptic battles of Revelation—the answer is not only yes, but profoundly yes, when understood through the lens of righteousness, justice, and divine purpose.

The Bible does not shy away from war. It places war within the framework of God's cosmic conflict against evil. In every dispensation—heaven, Old Testament Israel, the New Testament church, and the end-time—warfare is a tool used not to glorify violence but to confront rebellion, protect the innocent, and advance divine justice.

Warriors in God's Image

To be a soldier is to reflect, in some small and humble way, the attributes of the Divine Warrior:

- Justice – Fighting not for conquest, but to defend truth, liberty, and the oppressed.
- Discipline – Submitting to authority, living a life of order, self-control, and sacrifice.
- Protection – Laying down one's life, if necessary, to protect others—just as Christ did for us.
- Courage – Moving forward even when afraid, for the sake of a greater cause.

When a Christian enters the military, they are not necessarily stepping into a secular or pagan environment. They are stepping onto a battlefield where the values of Heaven can be lived out. Whether they wear the uniform of a Marine, a Sailor, a Soldier, or an Airman, they can wear the spiritual armor of God at the same time.

Divine Discipline and Earthly Purpose

Military service can be an unparalleled training ground for spiritual formation. It instills the virtues that God has called His people to live by:

- Obedience to authority – mirroring our submission to God.
- Sacrifice – echoing Christ's own.
- Brotherhood and unity – reflecting the body of Christ.

I have met some of the most devout believers not in pews, but in cammies. Men and women who pray with conviction in tents, who carry Bibles in their cargo pockets, and who sing praises beneath desert skies. Their faith is real, tested, and deeply rooted. They are living witnesses that faith and military service are not enemies—they are allies in the war for the soul.

A Heavenly Model for Earthly Warriors

When God created structure in His heavenly kingdom, He created rank, mission, and order. He appointed angels to command, to guard, to fight, and to minister. He called people like David, Joshua, Deborah, and Gideon to serve Him through military leadership. And in the end, He will bring peace to this universe through a final act of divine warfare.

If God's kingdom is organized like an army and He Himself is a warrior, how could military service be inherently unspiritual?

We do not glorify war—but we glorify the One who fights to redeem.

To Those Who Are Called

To you, reader—if you are wrestling with the question of whether God could possibly call someone like you into the military, hear this:

Yes, He can. And maybe He already is. Your service can be a sacred mission. Your training can shape your testimony. Your uniform can be a form of witness. Your calling can become a cause for salvation.

God doesn't need perfect vessels—He needs willing ones. The battlefield is not limited to spiritual metaphors. Sometimes, it looks like deployment orders. Sometimes, it sounds like boot camp. Sometimes, it takes you to places of death so that you might bring life.

"You therefore must endure hardship as a good soldier of Jesus Christ" (2 Timothy 2:3). This is not just an appeal for toughness—it's a divine reminder that military life, when surrendered to Christ, becomes a sacred assignment.

Final Thoughts

Are armed forces biblical? Absolutely—when understood through the right lens. War is not the enemy of faith. Sin is. And sometimes, to confront sin, evil, and injustice, God raises up warriors—men and women forged in fire, disciplined by trial, and marked by a cause greater than themselves.

If that's you—step forward. Your orders may come not just from a commanding officer, but from the Commander of Heaven's armies.

And if God is calling you to serve, know this: You are not abandoning your faith—you are answering a higher call.

CHAPTER 7

Biblical Combat in the Bible

THROUGHOUT SCRIPTURE, the people of God find themselves in battle—sometimes as instruments of judgment, other times as victims of oppression or defenders of justice. These battles, both physical and spiritual, reveal an unmistakable pattern: the Lord Himself is portrayed as a warrior who intervenes in human history to uphold righteousness. The Bible's record of warfare is not a glorification of violence but an exposition of divine sovereignty. God is neither indifferent to conflict nor dependent on human armies; He is the commander of heaven's hosts, fighting not for conquest but for covenant.

The earliest depictions of divine warfare appear in the Exodus narrative. Israel was a nation of slaves, without weapons or military training, pursued by one of the most formidable armies of the ancient world. Yet at the Red Sea, God displayed His supremacy not through human strategy but by His own hand. "The Lord will fight for you; you need only to be still" (Exod. 14:14, NIV). These words from Moses echo across generations, offering comfort to believers facing overwhelming odds. The Israelites were trapped between Pharaoh's chariots and the sea—an impossible situation by any human measure. But when Moses stretched out his staff, the waters divided, creating a path of deliverance. As the people crossed safely, the Egyptian army pursued and was swallowed by the collapsing waves.

This moment stands as one of Scripture's clearest demonstrations that victory belongs to the Lord. Israel did not win by weapon or wit; they were spectators of divine power. God's purpose in this event went beyond survival—it was revelation. He displayed to both Israel and Egypt that He alone is sovereign over creation, capable of delivering His people by means that defy reason.

For believers serving in the military, the lesson is profound: divine help is not limited by circumstance or technology. Modern warfare may rely on precision and intelligence, yet the ultimate outcome still rests in God's hands. Faith becomes the unseen armor that protects and sustains the heart. The believing warrior learns from the Red Sea that while human strength is necessary, dependence upon divine intervention is essential. In every age, God remains the defender of those who trust in Him.

God as the Divine Warrior

The Song of Moses that follows the Red Sea deliverance includes a striking declaration: "The Lord is a warrior; the Lord is His name" (Exod. 15:3, NIV). This statement introduces a theology that will echo through the Psalms and prophets—the idea of God as the divine warrior. In ancient culture, victory in battle was evidence of divine favor. Israel's God was unlike any pagan deity because His wars were always moral, not territorial. He fought not to expand borders but to establish justice and preserve His covenant promises.

In Psalm 24, David asks, "Who is this King of glory? The Lord strong and mighty, the Lord mighty in battle." The psalmist envisions God as the commander returning from war, victorious over chaos and evil. This imagery isn't metaphor alone; it defines a central truth about God's relationship with humanity. He does not remain distant from our struggles. When His people face oppression, He enters the battlefield on their behalf.

Throughout the Old Testament, God's interventions follow a pattern: He acts when His people are powerless. The armies of heaven engage when human resources have reached their limits. This divine participation reminds believers that combat, in the biblical sense, is never merely human; it is spiritual. The battles we fight on earth—whether physical wars, moral crises, or personal trials—mirror the cosmic struggle between good and evil.

For the believing warrior, this truth brings perspective. Serving in the military is not simply a career; it is participation in a long tradition of those who stand between order and chaos, reflecting the God who wages war for justice. A Christian in uniform embodies both strength and restraint, courage and compassion. The calling is sacred: to represent divine order in a world prone to destruction.

The Battle of Jericho — When Obedience Became a Weapon

When Israel finally reached the Promised Land, the fortified city of Jericho stood as their first obstacle. Militarily, the walls of Jericho were impenetrable. But God's instructions to Joshua were unusual—almost absurd by human standards. "March around the city once with all the armed men. Do this for six days. On the seventh day, march around the city seven times, with the priests blowing trumpets" (Josh. 6:3–4, NIV).

For an experienced warrior, such a strategy would seem illogical, even foolish. Yet Joshua obeyed completely. The people marched in silence, their obedience a form of worship. On the seventh day, the trumpets sounded, the people shouted, and the walls collapsed. Jericho fell not to siege weapons but to faith expressed through obedience.

This battle reveals that God's victories often defy human logic. Military might and tactical brilliance are valuable, but they cannot replace divine instruction. The fall of Jericho demonstrates that obedience is the greatest weapon in spiritual warfare. The believing warrior, therefore, must learn that success in any mission—whether combat or daily duty—depends more on listening to God than leaning on experience.

Joshua's soldiers were disciplined, but their discipline alone could not bring down walls. It was faith in God's word that made the impossible possible. The same principle governs the life of every Christian soldier today. Following orders matters—but following God's orders matters most.

In Jericho, Israel's faith became visible. Their march was a declaration of trust; their shout, a proclamation of victory before the walls fell. For the believer in uniform, this story becomes a metaphor for persistence in prayer, patience under pressure, and courage to act when God commands. Victory often arrives after obedience has been tested.

Faith in the Face of the Impossible — Lessons for the Believing Warrior

The biblical narrative repeatedly affirms that God partners with those willing to act in faith. When He fights for His people, He often does so through them. The Red Sea displayed God's direct intervention; Jericho revealed His cooperation with obedient faith. Together, these stories illustrate that divine warfare operates on two fronts: God's sovereignty and human submission.

For those who serve today, this partnership mirrors the dual nature of Christian service—active duty and spiritual duty. The soldier's obedience to command parallels the believer's submission to God's authority. Both require discipline, humility, and trust. The military profession, when exercised with moral integrity, becomes a living parable of spiritual truth.

When a Christian service member chooses to operate with honesty amid corruption, compassion amid cruelty, and restraint amid provocation, he becomes a reflection of God's righteous character in hostile territory. The soldier's battlefield may differ from Joshua's, but the principle remains: the Lord still fights through His people.

Jericho teaches that every believer is part of a divine campaign larger than himself. When a chaplain prays for a wounded sailor, when a medic risks life to save another, or when a commander refuses unjust orders, each becomes an echo of the same truth—God still wages war through human vessels who walk in obedience.

The stories of divine combat in Scripture are not relics of an ancient past; they are revelations of enduring principles. They teach that God values courage, obedience, and sacrificial service. Just as He called Joshua to march, David to fight, and Moses to lead, He still calls men and women today to stand between tyranny and peace.

To serve in the military, for the believer, is to step into the ancient rhythm of divine warfare—not for conquest or revenge, but for protection and justice. The believing warrior embodies God's own nature as defender and deliverer. Military service, when pursued with righteousness, becomes a sacred trust—an act of worship expressed through duty.

The God who parted seas and brought down walls still fights for His people. He simply does so now through those willing to wear the uniform with integrity and faith. Each act of courage, each decision guided by conscience, becomes a battlefield prayer—"Lord, fight through me."

In this way, the believer in uniform does not merely fight for a nation but participates in the timeless work of God—the preservation of life, the defense of justice, and the manifestation of divine love in a broken world.

Faith and Courage on the Battlefield

Scripture overflows with stories where human frailty meets divine strength. Each account reminds us that courage is not the absence of fear but the decision to trust God in the midst of it. Among these, two episodes stand out as portraits of what it means to fight—not merely with weapons—but with faith: David and Goliath, and Moses' struggle against Amalek.

David vs. Goliath — Faith Over Fear

The valley of Elah became the stage for one of history's most enduring battles. For forty days, the Philistine giant Goliath mocked Israel's army, his armor gleaming beneath the sun, his words cutting deeper than his sword. Soldiers trained for combat trembled; even King Saul, Israel's tallest and most capable warrior, stayed hidden in fear. Into that silence stepped David—a shepherd, not a soldier, carrying bread and obedience rather than armor and arrogance.

When David heard Goliath's defiance, something stirred within him that no weapon could ignite. "Who is this uncircumcised Philistine that he should defy the armies of the living God?" (1 Sam. 17:26, NIV). He saw what others missed: that this was not a contest of strength, but of faith. While the army measured height and weaponry, David measured covenant and calling.

Rejecting Saul's armor, David picked up five smooth stones from the brook. His confidence was not in his sling but in the Lord's faithfulness. As Goliath advanced, David shouted, "You come against me with sword and spear and javelin, but I come against you in the name of the Lord Almighty" (1 Sam. 17:45). One stone later, the battle was over.

David's victory redefined warfare. He proved that God's presence outweighs any military advantage. For believers in uniform, this story captures the heart of Christian service: courage is born when faith replaces fear. The believing warrior learns that moral courage—standing for truth, defending the weak, refusing injustice—is the modern battlefield of faith. The sling and stone may have changed into body armor and duty rosters, but the principle remains the same: victory belongs to those who fight in God's name, not their own.

David's story also reminds service members that preparation and faith coexist. He was skilled with his sling; faith did not negate training—it perfected it. The soldier who prays without practice is unprepared, and the one who trains without prayer is unprotected. David embodied both, making him the model of spiritual readiness.

His triumph foreshadows Christ, the greater Shepherd-Warrior, who would face humanity's giant—sin—and conquer it through faith and obedience. The same God who guided David's stone now guides every believer willing to stand where others retreat. In that sense, the modern warrior's courage becomes a continuation of the same faith that felled Goliath.

Moses, Aaron, and the Battle Against Amalek — Intercession and Teamwork

Soon after leaving Egypt, Israel met its first military test. The Amalekites ambushed the weary travelers at Rephidim. God commanded Moses to appoint Joshua as commander while Moses himself stood on a hill overlooking the battle, holding in his hand the staff that symbolized God's authority.

"As long as Moses held up his hands, the Israelites were winning, but whenever he lowered his hands, the Amalekites were winning" (Exod. 17:11, NIV). When fatigue overcame him, Aaron and Hur stood beside him, one on each side, supporting his arms until the sun set. Only then did Joshua's forces prevail.

This battle paints a profound image of cooperation between spiritual support and physical strength. Israel's victory required both the soldier's sword and the intercessor's prayer. If either failed, defeat followed. God designed the victory to remind His people that no warrior fights alone. Prayer fuels perseverance; community sustains courage.

In military terms, Moses, Aaron, and Hur represent the essential unity between command, support, and execution. Every mission depends on unseen hands that lift others in prayer, planning, or logistics. The modern Christian service member experiences the same reality when families, chaplains, and fellow believers hold them up before God. Behind every successful operation of justice lies a network of faith that refuses to let weary arms fall.

For the believing warrior, this passage reveals that combat is never purely physical. Every external battle mirrors an internal one—the struggle to keep faith raised high when exhaustion sets in. When Moses' arms trembled, Aaron and Hur's presence prevented defeat. Likewise, the Christian soldier must learn to lean on spiritual community. Isolation breeds defeat; fellowship brings endurance.

This story also clarifies that God values dependence more than independence. In a culture that glorifies self-sufficiency, the image of Moses being held up by others teaches humility. Even leaders need support. The best commanders, chaplains, and warriors are those who understand their need for God and for one another.

From a theological lens, the raised staff symbolized intercession—the lifting of the human will toward heaven. The battle below mirrored the spiritual warfare above. Just as Joshua's troops fought with swords, Moses fought with prayer. Both were acts of obedience; both were instruments in God's hands. This dual engagement—spiritual and physical—illustrates what it means to serve God in uniform. The soldier's duty and the saint's devotion are not separate; they are two sides of the same calling.

Lessons for the Believing Warrior

- Faith Must Lead the Fight. David's story reminds every believer that the greatest victories are won in the heart before they appear on the field. Faith gives direction to courage and prevents it from becoming recklessness. The believer's confidence must rest not in weapons, but in the righteousness of the cause and the presence of God within it.

- Community Sustains Victory. Moses' raised hands symbolize the shared nature of triumph. No battle is won alone—whether in ancient Israel or modern service. Prayer partners, families, and faith communities are the unseen warriors who ensure spiritual readiness. When one grows weary, others must hold him up.

- Obedience Is Greater Than Strength. Both David and Moses acted on God's instruction, not human logic. David refused armor; Moses raised a staff. Their obedience unlocked divine power. For the modern service member, obedience to moral conviction—even under pressure—is the mark of true strength.

- Intercession Is Warfare. Every time a chaplain prays on deck, a soldier bows before deployment, or a believer intercedes for peace, another unseen battle tilts toward victory. The power of prayer is not symbolic—it is strategic. It aligns heaven's will with earth's conflict.

These two biblical battles—one fought with a sling, the other with a staff—reveal that God honors both the warrior's courage and the intercessor's prayer. The military profession, when grounded in faith, becomes a field where divine purpose unfolds. The believing warrior stands as both protector and prayer partner, embodying David's courage and Moses' humility.

Serving in uniform allows Christians to live out this dual calling: to fight when justice demands it and to pray when mercy is needed. The soldier's posture—strong yet surrendered—reflects the balance of heaven's justice and love. God still works through disciplined hands and faithful hearts.

When believers serve, they do not glorify war; they glorify the God who brings peace through order, freedom through sacrifice, and redemption through obedience. Every mission becomes a modern echo of David's shout and Moses' uplifted hands—a declaration that the battle belongs to the Lord.

When Victory and Defeat Both Teach Faith

The Bible is not a book of unbroken triumphs. It is a record of real people—soldiers, servants, and leaders—who experienced both glorious victories and devastating defeats. In both, God revealed His character. In victory, He showed His strength; in defeat, He showed His holiness. For the believing warrior, these stories teach that winning every battle is not the measure of faithfulness—trusting God in every outcome is.

Gideon and the 300 — Strength in Small Numbers

When the Midianites terrorized Israel, God chose an unlikely leader: Gideon, a man hiding in fear, threshing wheat in a winepress to avoid enemy patrols. The angel of the Lord greeted him with irony and promise: "The Lord is with you, mighty warrior" (Judg. 6:12, NIV). Gideon's initial response—"Pardon me, Lord, but if the Lord is with us, why has all this happened to us?"—reflects the honest doubts of many believers called into service. Yet God saw courage where Gideon saw inadequacy.

Soon Gideon found himself assembling an army to face the Midianites, whose forces were "as numerous as locusts" (Judg. 7:12). But God intervened in an unexpected way. "You have too many men," the Lord told him, "I cannot deliver Midian into their hands, or Israel would boast against me" (7:2). Through a series of tests, God reduced the army from 32,000 to 300 men—less than one percent of the original force. Their weapons? Trumpets, torches, and clay jars.

At night, Gideon's small army surrounded the enemy camp. On his command, they shattered their jars, raised their torches, and blew their trumpets, shouting, "A sword for the Lord and for Gideon!" Panic erupted among the Midianites. "The Lord caused the men throughout the camp to turn on each other with their swords" (7:22). The vast enemy collapsed under confusion, and victory was secured without a single sword swung by Israel.

This victory was not about tactics—it was about trust. God deliberately weakened Israel to magnify His strength. Gideon's 300 were not chosen for skill but for submission. Their obedience made them instruments of divine strategy. The battle demonstrated that God's power is perfected in weakness, a truth echoed later by Paul in 2 Corinthians 12:9.

For today's Christian servicemember, Gideon's story is a reminder that effectiveness in God's army is not measured by numbers, rank, or recognition. A faithful few can change the course of history when God is their commander. In times of moral confusion or ethical challenge, standing firm—even when outnumbered—becomes the battlefield where faith is proven.

Gideon's courage did not begin on the field; it began in the heart. He tore down his father's altar to Baal before leading Israel into combat. His first act of warfare was spiritual, not physical. Likewise, for the believer in uniform, the true battle begins in private obedience before it is expressed in public duty. The uniform may distinguish soldiers by branch or rank, but faith distinguishes them by purpose.

Gideon's 300 remind us that the believing warrior is never defined by statistics but by surrender. When God chooses to use a small force to accomplish great deliverance, He proves again that the victory belongs to Him alone.

The Defeat at Ai — When Sin Undermines Strategy

If Gideon's story celebrates faith, the defeat at Ai exposes its opposite—disobedience. After the fall of Jericho, Israel was confident. Their morale was high; their enemies feared them. The next target, Ai, appeared small and weak. Joshua sent only a few thousand soldiers, certain the battle would be easy. But instead of victory, disaster struck. Israel fled in humiliation, and thirty-six men died in retreat. Joshua fell to the ground, crying out in confusion, "Why did You ever bring this people across the Jordan?" (Josh. 7:7).

The reason for defeat soon surfaced. A man named Achan had secretly taken gold, silver, and a robe from Jericho—items that God had declared "devoted to destruction." His private sin had national consequences. The Lord told Joshua, "Israel has sinned; they have violated My covenant" (7:11). Once the sin was confronted and purged, Israel went back to battle and won easily.

The defeat at Ai teaches a sobering lesson: no amount of military skill can compensate for moral compromise. Disobedience corrodes discipline; hidden sin weakens the entire body. For the believer in the military, integrity is not optional—it is armor. God does not bless victory built on corruption. A single act of dishonesty, injustice, or abuse of authority can undo the efforts of many. The contrast between Jericho and Ai is striking. At Jericho, the people obeyed and triumphed; at Ai, they presumed and failed. The difference was not tactical—it was spiritual. God uses both victory and defeat to refine His people, reminding them that success apart from righteousness is not success at all.

For Christian service members, this truth holds particular relevance. righteousness, outcomes over integrity. Yet the believing warrior understands that the presence of God, not the absence of obstacles, determines success. A moral defeat can wound deeper than a physical one, and repentance is the only road to restoration.

Lessons from Victory and Defeat

- *Faith Requires Vulnerability.* Like Gideon, we must allow God to strip away what makes us self-reliant. Strength in His service often begins with surrender. Whether it's a small team in the field or a believer standing alone for truth, God's power shines brightest through human weakness.
- *Disobedience Endangers the Mission.* Achan's hidden sin reminds us that holiness is not private property—it affects the community. For those in military service, integrity safeguards more than personal reputation; it protects collective trust and divine favor.
- *Victory Is Not Proof of God's Approval.* Israel's early success at Jericho tempted them to assume future victories were guaranteed. God sometimes allows setbacks to remind His people that dependence must be continuous, not occasional.
- *Loss Can Be a Lesson.* The believer who fails but repents learns more about grace than the one who wins without reflection. Defeat, in God's hands, becomes a classroom for humility.

The stories of Gideon and Ai illustrate that God works through warriors who value obedience above victory. Serving in the military offers believers a living stage for that truth. Just as Gideon learned that trust triumphs over numbers, today's Christian service members must rely on faith when facing moral and spiritual challenges.

The defeat at Ai reminds the believer that serving in uniform is not merely about courage—it is about character. A soldier's outward strength must mirror inward integrity. God honors those who serve with righteousness, humility, and dependence upon Him. When believers choose to serve, they step into a divine paradox: to fight for peace, to protect life through discipline, and to confront evil with humility. Each mission becomes an opportunity to demonstrate that God's kingdom advances not only through worship in sanctuaries but also through faithfulness on battlefields. The Bible's record of combat proves that God does not abandon warriors—He shapes them. Some He strengthens through victory; others He refines through loss. But all who trust Him learn the same lesson: obedience is the believer's greatest weapon, and faith is his surest defense.

The Ultimate Warrior: From Holy Battles to Holy Character

The story of combat in Scripture begins and ends with God Himself. Long before Joshua lifted a sword or David hurled a stone, the Bible presents the Lord as the commander of armies—the Yahweh Sabaoth, the "Lord of Hosts." From the Red Sea to Revelation, He is portrayed as a God who wars not out of malice but out of mercy, not to destroy indiscriminately but to preserve righteousness. In Exodus 15:3, Moses sings, "The Lord is a warrior; the Lord is His name." This revelation is not an afterthought; it is foundational to understanding divine justice. God fights when evil threatens His creation. His wars are never born from greed or revenge, but from holiness. While human conflicts often stem from pride, territory, or power, divine warfare flows from love—love that refuses to let evil reign unchecked.

The Psalms echo this reality repeatedly. "He trains my hands for battle; my arms can bend a bow of bronze" (Ps. 18:34, NIV).

David's confession captures both humility and empowerment: the warrior's skill is itself a gift from God. For the believer in uniform, this means that even tactical excellence, discipline, and readiness can be acts of stewardship. Military service becomes an extension of divine order—restoring peace through justice, strength, and moral courage.

When God Fought Alone

There were moments when God chose to act without human aid, reminding His people that their survival did not depend solely on their armies but on His sovereignty. One such event took place during the reign of King Jehoshaphat. Surrounded by a coalition of Moabites, Ammonites, and Edomites, Judah faced overwhelming odds. The king prayed publicly, "We do not know what to do, but our eyes are on You" (2 Chron. 20:12, NIV).

In response, God declared through His prophet: "Do not be afraid or discouraged because of this vast army. For the battle is not yours, but God's" (v. 15). The next day, the army marched—but not with weapons at the front. Instead, singers led the procession, praising God. As they worshiped, the enemy armies turned on each other in confusion until none remained.

Jehoshaphat's victory was not won by swords but by surrender. The frontline of battle became a choir loft, and the weapons of war were hymns of faith. God fought alone, proving again that His strength does not depend on human force. For believers serving in the military, this passage reminds us that even in an institution of might, dependence on God remains the greatest defense. The believing warrior must remember that success is not measured by firepower but by faithfulness. God's power often manifests most clearly when His people acknowledge their limits.

Jehoshaphat's story also redefines courage. Courage is not merely running toward danger—it is worshiping in the face of it. The believer who kneels before battle demonstrates the deepest form of strength: submission to the ultimate Commander.

Christ and the Transformation of Warfare

When Christ entered the world, He did not erase the imagery of warfare; He transformed it. The Messiah came not with armies but with authority, not with legions of angels but with the cross. His battlefield was not the plains of Megiddo but the hill of Calvary. There, He waged the final and greatest war—not against nations, but against sin and death.

The Apostle Paul captured this transformation when he wrote, "For our struggle is not against flesh and blood, but against the rulers, against the authorities, against the powers of this dark world" (Eph. 6:12, NIV). The weapons of this new warfare were no longer swords or shields but truth, righteousness, faith, and prayer. Yet the warrior imagery remained —because spiritual conflict is as real as physical combat.

In Revelation 19, the returning Christ is portrayed as a conquering warrior: "With justice He judges and wages war… On His robe and on His thigh He has this name written: KING OF KINGS AND LORD OF LORDS." The same Jesus who taught peace also executes justice. This balance between mercy and judgment defines the heart of divine warfare.

For the believer in the military, Christ's example brings clarity: service is not about domination but about redemption. The uniform becomes a symbol not only of national defense but of divine order—of standing in the gap between chaos and peace. The believing warrior serves as a living reminder that peace without justice is fragile, and justice without love is incomplete.

To fight righteously is to imitate the Savior who fought evil not to destroy humanity but to deliver it. His battle cry was not "conquer" but "forgive." And yet, even forgiveness was a form of warfare—one that disarmed the powers of hell.

The Warrior Spirit of the Believer

Though modern believers may not march into battlefields of blood and dust, they still fight daily wars—against despair, injustice, temptation, and moral compromise. The soldier of faith carries both sword and spirit, discipline and devotion. Every believer, whether in uniform or civilian life, is enlisted in this spiritual army, called to resist evil and defend what is good.

For those who serve in the armed forces, this calling carries a double weight. Their profession embodies both the physical defense of freedom and the spiritual defense of truth. The same courage required to face a visible enemy is needed to confront the invisible ones: doubt, fear, and moral decay.

The Christian warrior understands that training the body without training the soul leads to imbalance. As the Apostle Paul wrote, "Physical training is of some value, but godliness has value for all things" (1 Tim. 4:8). The believer in uniform must therefore cultivate both readiness and righteousness. Faith does not weaken a soldier—it completes him.

In every deployment, drill, or mission, the believing warrior stands as a testimony that God still calls His people into places of conflict, not to perpetuate violence but to restrain it. Their service is an act of stewardship—preserving life, protecting the innocent, and modeling integrity in systems often marked by corruption or compromise. In this way, the believer's service mirrors God's own mission: bringing order out of chaos.

From Exodus to Revelation, Scripture portrays a consistent truth—God is not a pacifist in the face of evil. He is a God of peace who wars for righteousness. When believers enter military service with prayerful hearts and pure motives, they are not abandoning their faith; they are extending it into one of the world's most difficult arenas.

Just as God raised up Gideon, Joshua, David, and countless others to stand between destruction and deliverance, He still calls men and women of faith to serve with courage and compassion. Military service becomes an act of discipleship when guided by moral conviction and love for humanity.

The modern believer who puts on the uniform does so not to glorify conflict but to embody Christ's justice in a broken world. Each mission, each sacrifice, and each act of service echoes the ancient truth sung by Moses: "The Lord is a warrior; the Lord is His name."

In the end, every battle—spiritual or physical—reminds us that victory belongs not to those with the sharpest sword but to those with the purest heart. The believing warrior, like his biblical forebears, fights not to destroy, but to protect, to restore, and to reveal that even amid warfare, God is still love.

CHAPTER 8

Military vs. Spiritual Training

LIFE IS A BATTLEGROUND. IT REQUIRES resilience, discipline, and an unwavering commitment to endure through challenges. Every Marine understands that the transformation from civilian to warrior does not happen overnight. It is forged through weeks of relentless training, hardship, and challenges that push recruits beyond their perceived limits.

This experience mirrors the spiritual journey of a believer. The Christian life is not a path of ease and comfort—it is a rigorous course that demands perseverance, faith, and obedience. Just as Marines must endure boot camp to earn the title, believers must navigate the trials of life to receive their ultimate reward when Christ returns.

In this chapter, I will share my personal experience of Marine Corps boot camp in 2003, drawing a parallel between the intense training that shaped me as a warrior and the spiritual preparation that shapes us as believers. The 13 weeks of training that transformed me from a struggling young man into a Marine taught me lessons that still impact my faith today. Just as we had to be stripped of our old selves to be rebuilt as Marines, so must believers be stripped of their sinful nature to be transformed into the image of Christ.

The Making of a Marine: The 13-Week Transformation

Boot camp is designed to transform civilians into warriors. It is a process that requires recruits to break away from their former selves, be rebuilt from the ground up, and emerge as disciplined Marines. This transformation happens in three key phases—a process that also mirrors the journey of faith.

Phase 1: The Breaking Down (Weeks 1-4)

The moment we arrived at Parris Island, we were no longer in control of our own lives. The drill instructors made sure of that. The second we stepped off the bus, chaos erupted. Drill Instructors (DIs) screamed orders, demanded absolute obedience, and made it clear that our old identities were gone.

I still remember standing on the famous yellow footprints, hearing the commands that would dictate my next three months. Everything I had known—my name, my independence, my sense of self—was being stripped away. For the first four weeks, we weren't allowed to refer to ourselves as "I" or "me." Instead, we had to say, "This recruit requests permission to speak, sir!"

The physical training was relentless. The drill instructors controlled every aspect of our lives. We were forced into early morning runs, hours of push-ups, obstacle courses, and extreme sleep deprivation. The goal was not just to train our bodies—it was to reshape our minds, to erase selfish thinking, and instill a new mindset of discipline and obedience.

Spiritual Parallel: The Breaking of the Old Self

The first step in the Christian journey is the breaking down of the old self. The Apostle Paul explains it clearly in 2 Corinthians 5:17: "If anyone is in Christ, he is a new creation. The old has passed away; behold, the new has come."

Just as boot camp breaks down a recruit's old habits and identity, so too does faith require us to let go of our past. Coming to Christ is not about adding religion to our lives—it is about transformation. Our sinful habits, pride, and selfish desires must be stripped away, and we must submit ourselves fully to God's process of renewal.

Jesus Himself emphasized this in Luke 9:23: "Whoever wants to be my disciple must deny themselves and take up their cross daily and follow me." Denying oneself is not easy. It means surrendering personal ambitions and embracing God's purpose, much like a recruit must submit to the authority of the drill instructors.

Phase 2: The Building Up (Weeks 5-9)

After breaking us down, the Marine Corps began building us back up. The early weeks had stripped away our individuality, but now we were learning how to function as a unit. Every lesson reinforced the idea that we were no longer individuals—we were a team.

We spent weeks perfecting our marksmanship skills, learning battlefield strategies, and training for combat scenarios. Our bodies became stronger, our endurance greater. Yet, the biggest transformation happened in our minds. We were no longer thinking of ourselves—we were thinking of the mission and those fighting beside us.

One of the most critical lessons we learned was trusting our unit. No Marine fights alone. Our survival and success depended on absolute trust in one another.

Spiritual Parallel: Training in Righteousness

Just as Marines must be trained for battle, so must Christians be trained for spiritual warfare. 1 Timothy 4:7-8 says: "Train yourself to be godly. For physical training is of some value, but godliness has value for all things, holding promise for both the present life and the life to come."

Spiritual growth is not accidental—it requires discipline, study, and community. A Christian who isolates themselves is like a soldier without a unit—vulnerable to attack. The Bible repeatedly emphasizes the importance of fellowship in spiritual training: "As iron sharpens iron, so one person sharpens another." (Proverbs 27:17)

Just as Marines train alongside their brothers and sisters, Christians must surround themselves with fellow believers who challenge and strengthen their faith.

Phase 3: The Final Test (Weeks 10-13)

After months of rigorous training, we faced The Crucible, a 54-hour test that would determine whether we were worthy of the title Marine. This was the most grueling experience of boot camp, designed to push us beyond our mental, physical, and emotional limits. During The Crucible, we carried 80-pound packs, ran miles without food or sleep, completed combat exercises, and were forced to rely on our training under extreme pressure.

When we finally reached the final march known as the Reaper, exhaustion and hunger made every step feel impossible. But at the top of that final hill, our Drill Instructors presented us with the Eagle, Globe, and Anchor—the symbol of the United States Marine Corps.

In that moment, all the suffering, all the exhaustion, all the struggle—it was all worth it. We had earned the title Marine.

Spiritual Parallel: The Ultimate Reward

Life itself is a spiritual Crucible. It is filled with trials, hardships, and challenges that test our faith. But if we endure, the reward is eternal. Paul compares the Christian life to a race that must be finished: "I have fought the good fight, I have finished the race, I have kept the faith. Now there is in store for me the crown of righteousness, which the Lord, the righteous Judge, will award to me on that day." (2 Timothy 4:7-8).

Just as The Crucible tests recruits to see if they are ready to be Marines, the trials of life test believers to see if they are faithful to Christ. Our reward is not a Marine emblem, but something infinitely greater: "Be faithful, even to the point of death, and I will give you life as your victor's crown." (Revelation 2:10)

The Journey of Endurance

Looking back, I now see that my experience in boot camp was not just about becoming a Marine—it was about preparing for life itself.

- It taught me discipline—just as faith requires daily commitment.
- It taught me brotherhood—just as faith requires fellowship.
- It taught me perseverance—just as faith requires enduring to the end.

For anyone struggling in their faith, remember:

1. God is training you – The trials you face are preparing you for something greater.
2. You are not alone – Just as Marines rely on their unit, believers must rely on their spiritual family.
3. The reward is worth it – If we remain faithful, Christ will one day return to crown us with eternal victory.

Just as I earned the title United States Marine, we will one day receive our final reward when Jesus returns, welcoming us into the Kingdom of Heaven. Until then, we press on—enduring the spiritual training of life with faith, courage, and perseverance.

My journey wasn't clean. It wasn't linear. But it was holy. And yours can be too.

I want to speak directly to someone reading this who thinks, "It's too late for me." It's not. If God waited for me, He will wait for you. If He redirected Moses at a burning bush, He can meet you in your living room, your barracks, your car. If He used my failures to mold me into a minister,

He can use yours to birth something beautiful.

You don't have to be perfect to be called. You just have to be willing.

God's Faithfulness, Not Mine

At the heart of this chapter is not my discipline, my loyalty, or my eventual obedience. At the heart of this story is God's faithfulness. It's His willingness to wait. His commitment to pursue. His refusal to let go, even when I did. It's His relentless grace that reached through my rebellion, my confusion, and even my pride to remind me that He finishes what He starts.

So when I wear my uniform today—whether it's clerical or military—it's not a badge of my achievements. It's a symbol of His patience. His mercy. His vision. I am not a pastor because I earned it. I am a pastor because God kept His promise.

And He still does.

Physical Fitness Training

I remember the first time I realized I wasn't ready for war. It didn't happen in a combat zone with rounds cracking overhead. It didn't come during an ambush or when I stared down the enemy through iron sights. No — my reckoning came far earlier, in a moment far less cinematic. It was just me, a scorching concrete parade deck, and the unmistakable weight of my own unpreparedness.

I had just stepped off the bus at Marine Corps Recruit Depot. The welcome was as warm as expected — thunderous shouts, relentless orders, and a flurry of motion that stripped us of every piece of civilian identity we had clung to. My name was gone, replaced with "Recruit." My clothes were traded for standard-issue cammies. My hair disappeared in a matter of minutes. My sense of comfort and control vanished with it. Everything I knew was replaced with a system that didn't care about who I was — only about who I would become.

The Pain of the First Run

That first night, we were introduced to the Initial Strength Test — the IST. It was our first physical assessment, a trial by fire to determine if we had what it took to survive the months ahead. I stood in line watching recruits attempt pull-ups, perform crunches, and run with every ounce of effort they had left. Some looked sharp, others already looked like they regretted stepping off that bus.

I thought I'd do fine. I had lifted weights back home. Ran on the treadmill. Did push-ups when I remembered. I thought I was in shape.

I was wrong.

When it was my turn, I grabbed the pull-up bar and realized my arms felt heavier than usual. The drill instructor (DI) barked, "Begin!" I managed nine pull-ups — barely clearing the minimum requirement. My arms burned. My pride stung worse. The crunches were manageable, but not impressive. I hit just over eighty in two minutes. Then came the run — a 1.5-mile sprint that felt more like a marathon wrapped in humiliation.

I took off too fast, trying to prove something. Within the first half-mile, my breathing was ragged. My stride shortened. Marines half my size breezed past me with a steady rhythm. One of the female recruits from another platoon shouted cadence as she passed me, her voice steady, her steps sharp. I was gasping. My vision tunneled. I crossed the finish line with a time that barely passed. My legs trembled, my lungs ached, and my ego was in shambles.

That night, lying in a top rack under the low hum of fluorescent lights, I stared at the ceiling and wrestled with something deeper than physical fatigue. My body ached, yes. But what bothered me most was the realization that I had shown up unprepared — not just physically, but spiritually too.

I had coasted into boot camp on surface-level strength — both bodily and soul-deep. I had prayed before leaving home. Asked God to watch over me. Said the right words. But now, in this crucible of fire and sweat, I realized how shallow my preparation had been. I had underestimated what would be required of me. And that moment, lying flat on my back with sore muscles and bruised pride, became a turning point. I didn't just want to pass — I wanted to be ready. I wanted to be fit.

The Marine Corps didn't need me to be impressive. It needed me to be dependable. It needed me to endure.

And the same is true for the Kingdom of God.

1 Timothy 4:8 says, "For physical training is of some value, but godliness has value for all things, holding promise for both the present life and the life to come." That verse took on new meaning for me that night. Physical strength matters in the Corps — because lives depend on it. But spiritual strength? That's the lifeline of eternity. If I couldn't carry my own body in a run, how would I carry my brothers through combat? And more than that — if I couldn't carry my own cross, how could I ever help someone else shoulder theirs?

The IST was a mirror, and what it reflected back wasn't just physical weakness. It was a spiritual wake-up call. A warning that I had treated preparation as a hobby instead of a lifestyle. I had been casual with things that should've commanded my full attention.

Discipline in the Corps is not a suggestion — it is a survival mechanism. Likewise, spiritual discipline is not a luxury — it is the difference between growth and spiritual stagnation, between standing firm and falling away when trials hit.

That night, I didn't pray for boot camp to be easier. I didn't ask God to lighten the load. I asked Him to shape me into someone who could carry it. I asked Him to tear down the parts of me that had grown lazy, soft, or entitled — the parts that had mistaken comfort for strength and convenience for commitment.

I asked Him to make me ready — not just for the Marine Corps, but for the war that rages in every heart and in every moment of life.

If boot camp was going to hurt, I wanted it to hurt in the right direction. If I was going to sweat, I wanted it to be the sweat that builds something solid. If I was going to be broken down, then let it be in a way that laid the foundation for something stronger — something that wouldn't collapse when the storms came.

Romans 5:3-4 came to life in those first few days. "Not only so, but we also glory in our sufferings, because we know that suffering produces perseverance; perseverance, character; and character, hope." I began to understand that suffering wasn't the enemy — slackness was. Pain wasn't the enemy — complacency was.

That was the beginning of a mindset shift. I started to see every pull-up, every sprint, every exhausted gasp not as punishment — but as preparation. Not as misery — but as molding.

Physical training demanded everything. And I began to realize that the Christian life was no different. The stakes were even higher.

The Marine Corps was preparing me to defend my country. God was preparing me to defend something eternal. And both demanded that I stop living as if effort was optional.

That first run didn't just break my stride — it broke my illusion.

And in that breaking, God began to build something I could never construct on my own.

I joined the Marine Corps. I had only been living in the state of Maryland for about five years, which meant that I had only been learning the English language for that amount of time and was not as proficient as I would have wanted to. I still wrestled with the culture shock and getting used to the American ways.

Even though I had finished three years of high school and I could understand parts of a conversation with someone, yet I was still not in the condition to join a militia force. Military was not an option for me. My raw English disqualified me to be a military man, lest a Marine. I could barely understand the people speaking on the movies lest I would understand a command from a drill instructor, or worse, a call for help in the battlefield.

Boot Campt Breakdowns

Every day in boot camp felt like a test. Not just of muscle, but of mind. Of will. Of spirit. There was no escape from the intensity — no room for excuses. From the moment reveille shattered the silence before sunrise, we were moving. Running. Carrying. Holding position. Getting screamed at. Sweating through our uniforms before most of America had even opened their eyes. What surprised me wasn't how physically hard it was — it was how relentless it was. There was no pause, no reset. You just kept going.

The idea was simple: break you down, then build you back up. But the Marine Corps doesn't build from scraps. They want foundation. They want raw potential that refuses to quit. I started to see the drill instructors differently — not just as tormentors, but as heat that forged something stronger from something soft. They didn't care how fast you could run on day one. They cared if you could still run on week ten with a rifle, a full pack, and twenty miles of dirt under your boots. They cared about endurance.

And nowhere was that more clear than during our Physical Fitness Test — the PFT.

Unlike the IST, the PFT was the true benchmark. It tested not just if you could survive basic training, but if you could perform as a Marine should in a real-world combat scenario. The standards were exact: pull-ups, crunches, and a 3-mile run. For a max score, you had to complete 23 pull-ups, perform 115 crunches in under two minutes, and run three miles in 18 minutes flat. There were age and gender brackets, of course, but the expectations were universal: show up ready, or don't show up at all.

The lead-up to that test pushed me harder than I'd ever pushed myself before. Pull-ups in the sand pit until my arms gave out. Crunches until my abs locked up. Runs before breakfast, after dinner, during firewatch rotations. There was no escaping the expectation that we would be better than we were yesterday. I started to understand something profound — the body grows when you stress it consistently. The soul is no different.

Christians love comfort, but they forget that growth comes from resistance. It's the trial that refines faith. It's pressure that reveals conviction. Without suffering, endurance can't be built. And without endurance, you won't finish the race — not in war, and not in faith.

Hebrews 12:1 commands us: "Let us run with endurance the race marked out for us." That sounds poetic until your lungs are burning and your legs scream at you to quit. Endurance sounds noble until you're in a spiritual season where the answers don't come, where the temptations increase, where the trials keep stacking up and God seems silent.

But that's the test. And it's always pass or fail.

I watched good men in boot camp fall apart not because they weren't physically strong — but because they didn't prepare for the pain. They didn't expect the test to be so relentless. One recruit was built like a statue. He looked like he walked off a bodybuilding poster — but he couldn't run. He couldn't carry weight over distance. His body was strong, but not conditioned. He failed the run twice, and when it became clear he couldn't pass, he was dropped to another platoon. We never saw him again.

He taught me something without saying a word. You can look like a warrior on the outside and still be weak where it counts.
That lesson haunted me for days. How many Christians walk around with a strong appearance — well-versed in scripture, dressed for Sunday, fluent in the language of church — but have never trained their soul for war? What happens when they're hit with grief? When God says "wait" instead of "go"? When temptation knocks for the thousandth time?

They crumble. They fade. They fall back — not because they didn't believe, but because they didn't prepare. They mistook knowledge for strength. But strength, in both war and faith, is measured in the fight — not the facade.

Boot camp didn't just teach me how to do push-ups. It taught me to stop trusting the illusion of readiness. It taught me that I couldn't coast into victory on yesterday's effort. Every day was a choice: push through or be pushed back. Grow or get dropped.

Our drill instructors were ruthless, but not without reason. One morning, after a brutal PT session in the rain, one of them stood before us, soaked and serious. He said, "If you collapse, someone else dies. If you slow down, your fire team suffers. If you fall out, your buddy carries your pack. You don't train for you. You train so someone else lives."

That hit differently.

I realized that same truth exists in our walk with Christ. Your spiritual fitness isn't just for you. When you're strong in faith, you're better equipped to encourage your brother. When you're disciplined in the Word, you're quicker to discern lies. When you train yourself in prayer, you're ready to intercede for others when they're too broken to speak.

Galatians 6:2 says, "Carry each other's burdens, and in this way you will fulfill the law of Christ." But you can't carry what you're too weak to lift. That includes grief, doubt, and accountability. If we want to be warriors in the Kingdom, we have to train like it.

The PFT pushed me beyond what I thought I could endure. I didn't get a perfect score, but I passed with everything I had. And for the first time, I didn't care about looking impressive. I cared about being ready. Not just for combat — but for the spiritual battles that would one day come, unannounced and unrelenting.

Boot camp broke me. But in the breaking, I started to see what God was trying to build.

Combat Conditioning and the Spiritual War

After graduation, I thought the hard part was over. I had made it through boot camp. I had earned the title. I had stood tall during the Eagle, Globe, and Anchor ceremony with tears in my eyes and a fire in my chest. But as any Marine will tell you, earning the title is just the beginning. You don't become a warfighter at graduation — you become one in Marine Combat Training.

MCT is where the mask of ceremony is stripped away and replaced with the reality of combat. No more motivational speeches. No more rehearsed inspections. Just weapons, dirt, fatigue, and learning how to survive. How to move under fire. How to think when your mind is foggy with exhaustion. How to keep your buddy alive when you're too tired to keep yourself standing.

You're trained to operate under chaos. Under weight. Under threat. And perhaps most importantly — under pressure.

One of the most revealing experiences of that period was the Combat Fitness Test, or CFT. Unlike the PFT, which tested individual endurance and strength, the CFT was all about combat-readiness. It was fast. It was violent. It was real.

Here's what it looked like: first, an 880-yard sprint — not just a run, but a dead sprint in boots and uniform, timed to the second. Then, the ammo can lifts — hoisting a 30-pound can repeatedly above your head for two minutes straight. You're expected to push out as many reps as possible. Your shoulders feel like fire. Your breath cuts short. Your will is tested in every rep.

Then comes the worst part: the maneuver-under-fire course. You crawl under simulated barbed wire. You drag your buddy in a fireman's carry. You sprint while dodging imaginary obstacles, lunge through cones, throw a mock grenade, and finish with a casualty evacuation — dragging a fellow Marine the length of a football field while gassed out and dizzy.

It's not meant to feel achievable. It's meant to mimic combat — and combat doesn't care if you're tired.

During one iteration, I nearly collapsed halfway through the ammo can lifts. My shoulders locked up. I grunted through clenched teeth. I wanted to stop. Everything in me screamed to let go. But beside me, another Marine shouted over the noise, "Let's go, brother! One more! Don't stop now!"

That one voice sparked something in me. I pushed. One more lift. Then another. I finished on empty. He clapped my back after we dropped the cans. "That's what we do," he said. "We carry each other when we're about to break."

Those words burned into my soul.

That's what we do.

And I couldn't help but think — Why doesn't the Church look like this more often?

In that moment, the spiritual application became as vivid as the ache in my body. We were trained to never leave a Marine behind, to bear one another's burdens, to step in when someone else falls short — not to judge them, not to leave them behind, but to carry them.

In spiritual warfare, it's no different. You see a brother slipping into despair, you grab him. You see a sister faltering in her faith, you step in. You don't wait for them to hit rock bottom — you carry their pack. You speak life when they've gone silent. You pray when they can't. You intercede. You move under fire.

But too often in the Church, we treat spiritual weakness as shameful — something to hide. We let people bleed out in silence. We isolate them instead of stepping into the fight beside them. We've confused spiritual community with performance, and forgotten that combat is communal.

Ephesians 6 tells us about the Armor of God — the belt of truth, the breastplate of righteousness, the helmet of salvation, the sword of the Spirit. It's a powerful image. But here's the reality: all the armor in the world is useless if the person underneath it hasn't trained for war. You can dress a recruit in body armor and give him a rifle, but if he's never been under pressure — if he's never run through rounds, never felt the sting of fatigue, never learned to maneuver when his brain is fogged and his body is failing — he's a liability. Not an asset.

The same is true for believers. You can quote scripture. You can wear the title. You can look like a Christian. But if you're not spiritually conditioned — if you haven't learned how to worship when the enemy is loud, how to trust when the answers are delayed, how to obey when it costs you — you'll fall in the fight.

Spiritual training isn't glamorous. It doesn't happen on Sunday morning in a perfect row of pews. It happens in the early mornings when you open your Bible even though you're tired. It happens in late nights when you choose to pray instead of scroll. It happens when you forgive, when you stay, when you sacrifice — all without applause.

One night during MCT, we did an overnight movement through the brush, full gear on, moving by compass and red light. We were tired. Silent. Focused. The corporal leading us stopped at one point and whispered, "Don't forget: when it gets dark, that's when the enemy moves."

That stuck with me.

The enemy moves in darkness — and not just in the field. In life too. In the quiet corners of your mind. In the unchecked anger. In the ignored conviction. In the weariness. That's when the fight turns real. And if you're not trained — not spiritually conditioned — you won't see it coming until it's too late.

I came out of MCT stronger, leaner, more alert. But I also came out more spiritually awake. Not because I had memorized more verses, but because I had finally started to live what I claimed to believe.

Every sprint, every lift, every crawl under concertina wire showed me something about my faith: if it's not strong under stress, it's not strong at all.

Faith that folds under fire is just theory. But faith that pushes through — faith that drags your soul forward when everything hurts — that's real. That's what it means to be fit for the fight.

Fleet Life: Real Battles Begin

I thought once I reached the Fleet, the real fight would be over. I had survived boot camp. Conquered the Crucible. Dragged Marines through the dust at MCT. I'd proven I could endure pain, perform under pressure, and finish strong. I expected things to get easier.

But the Fleet is where I learned something far more dangerous than pain — I learned about complacency.

In boot camp, we were constantly pushed. If you slowed down, someone corrected you. If you slacked, it was noticed immediately. Discipline was imposed. Accountability was automatic. But once I got to the Fleet, that external pressure faded. No one forced you to go the extra mile. No one checked if you were training hard unless you wanted to be ready. You had to choose discipline — or watch yourself drift.

And that's where many Marines started falling apart. Not in war. Not in the field. But in barracks rooms, alone, comfortable, soft. I saw men who had once run six-minute miles now winded after a single lap. I watched Marines gain thirty pounds in three months. Not because they were lazy — but because the urgency was gone.

There were no drill instructors yelling in their faces. No test on the calendar. And without a visible threat, they stopped preparing for one.

The Fleet taught me this: the absence of war doesn't mean the war is over — it just means you can't see it coming yet.

And I realized the same is true for the Christian life.

It's easy to be disciplined when you're in a storm. When your back is against the wall. When you're fighting for your marriage, your sanity, your purpose. You cling to the Word because it's your only anchor. You hit your knees because nothing else makes sense. In the fire, you train hard.

But what about when things are good?

When the bills are paid. When no one's sick. When the prayers are being answered. When the spiritual high wears off, and there's no immediate threat — do you still train?

Because that's where most Christians fall.

Not in the fight — but in the comfort.

One of the hardest hikes I ever went on was during a field op in California. The sun was merciless. The packs were heavy. We moved in formation across a brutal ridge line. Halfway through, one of our guys collapsed. He'd been skipping PT, assuming he was strong enough to get by. He wasn't. He hadn't trained. And when the weight hit, he went down.

We redistributed his gear and carried him out.

That hike became a living parable. A visual sermon. Because in the same way, I've seen believers collapse under pressure — not because they didn't love God, but because they hadn't trained with Him. They didn't maintain their spiritual conditioning. They assumed yesterday's strength would carry them through today's test. But faith doesn't store like canned goods. It has to be kept fresh — daily.

Jesus said in Luke 21:34, "Be careful, or your hearts will be weighed down with carousing, drunkenness and the anxieties of life, and that day will close on you suddenly like a trap."

The enemy doesn't always show up with horns and fire. Sometimes he shows up as convenience. Distraction. Success. Ease. And when we drop our spiritual guard, we don't notice how slow we've become, how soft, how unaware.

Complacency is quiet. That's why it's so deadly.

There were days in the Fleet where I had to drag myself out of bed and go run, not because anyone told me to — but because I remembered what it felt like to be unready. To be winded on that first run. To be the weakest link.

And in the same way, I've learned to wake up early and pray. Not because God will punish me if I don't — but because I remember what it felt like to face a trial unprepared. To be dry in the spirit. To be spiritually flabby when the enemy came knocking.

1 Corinthians 10:12 warns us: "So, if you think you are standing firm, be careful that you don't fall!"

The Fleet will fool you. Life will fool you. It will lull you into a false sense of security — that you've arrived, that you've made it, that you no longer have to train. But the truth is, spiritual fitness is never finished. The moment you stop moving forward, you start drifting back.

Discipline has to become personal.

Not forced. Not conditional. Not driven by fear. But by conviction — that if I stop training, I become a risk. Not just to myself, but to others. If I'm not strong, someone else might carry my load. If I'm not alert, someone near me might take the hit.

In the Corps, we train so we don't fail each other in combat. In the Kingdom, we train so we don't fail each other in trials.

And that's what this taught me — that the real battles often don't come with bullets or explosions. They come quietly. Slowly. Through the erosion of discipline and the slow creep of comfort. They come when we forget that there is no offseason in warfare — only silent battlefields.

Unfit to Fight – A Risk to Mission

There's something terrifying about knowing you're not ready. Not just "uncomfortable," not just "out of shape," but fundamentally unfit. I've seen it in a fellow Marine's eyes before a hike he hadn't trained for — that haunted look that says, I might not make it. It's not just the fear of failure that sets in; it's the fear of being a burden to the team. Of knowing that if your body quits, someone else will have to carry your weight. That fear sits in your gut like a rock, heavy and silent.

But even more terrifying than physical unreadiness is spiritual unreadiness. Because in spiritual warfare, the consequences aren't soreness or delayed graduation — they're devastation. They're broken homes, derailed callings, compromised integrity, and people falling under the weight of trials they were never conditioned to withstand. I've seen too many Christians drop out of the fight not because they didn't love God, but because they never trained for what was coming.

We train in the Marine Corps so we don't fail the mission. So we don't cost someone else their life. That's not just a motivational line — that's a battlefield reality. You carry your weight so others don't die under it. The same is true in the Kingdom of God. Your spiritual discipline isn't about looking holy; it's about being dependable when the enemy starts advancing. Because make no mistake — he always does.

I'll never forget the hike that taught me this. We were ten miles deep, full gear, thick mud, brutal heat. One of our guys started fading fast — the kind of slow fade you notice too late. He didn't speak up. Maybe he was too proud. Maybe he thought he could push through. But halfway through the ridgeline, he collapsed. Heat stroke. Dehydration. Total shutdown. We had to stop the formation, redistribute his pack, call a corpsman, and carry him the rest of the way. He didn't intend to be a burden. He didn't want to fall out. But he hadn't prepared, and the weight caught up to him.

That night, as we debriefed, our squad leader said something I'll never forget. "He wasn't just a danger to himself. He became a risk to all of us." That sentence hit me in the chest. It wasn't said with hate. It was said with sobering honesty. In combat, the unfit Marine doesn't just struggle — he puts others at risk. He slows the team. He breaks the formation. He creates openings for the enemy.

And in that moment, the Spirit of God whispered something equally sobering to me: So does an unfit Christian.

We don't like to think that way. We want church to be soft, welcoming, gentle. And it should be full of grace. But it also has to be full of grit. Because we are at war — not with flesh and blood, but with principalities and powers. And in war, the spiritually lazy become liabilities. The spiritually untrained leave gaps in the wall.   The spiritually undisciplined are easy to deceive, easy to distract, easy to disarm.

There is a weight that comes with being a warrior in the Kingdom. A

CHAPTER 8: MILITARY VS SPIRITUAL TRAINING

weight that says, "You carry more than your own life." Your prayers matter. Your endurance matters. Your integrity matters. Your strength or weakness ripples into the people God has assigned to you. And when you fall, they may feel the hit.

That's not condemnation. That's responsibility.

2 Timothy 2:3 says, "Endure hardship with us like a good soldier of Christ Jesus." Not a soft believer. Not a casual churchgoer. A soldier. One who trains. One who obeys. One who endures. And I'll be honest — I didn't understand that verse until I started living it in the dirt, in the pain, under weight I didn't think I could carry. But the more I trained, the more I realized: hardship doesn't come to crush the soldier. It comes to prove him.

And if you don't train for it, it'll break you.

You don't become a warrior by attending one church service. You don't become spiritually fit by quoting a few verses or posting about God online. You become fit when you live this thing. When you wake up early to seek God before the day hits you. When you say no to temptation that no one else would know about. When you carry someone else's pain in prayer until they feel light again. When you press into obedience even when it costs you everything.

There were mornings in the Fleet where I had to lace up my boots before the sun rose, step out into the cold, and run — not because I felt like it, but because I had a responsibility to stay ready. It wasn't about me anymore. It was about the man next to me who might need me in a firefight. About the mission that could drop at any time. And every day I stayed sharp, I reminded myself: readiness isn't seasonal.   It's sacred.

I carry that into my walk with Christ now. I read the Word daily not because I want a check mark, but because I don't want to die spiritually out of shape. I pray because my mind needs clarity before the day clouds it. I fast because I want my flesh to know who's in charge. I fight for purity and obedience not out of fear of punishment — but because I've seen what happens when good men don't train.

We need warriors who are fit. Not flashy. Not famous. Fit.

Warriors who can discern lies because they've been in the Word.

Warriors who know how to suffer without quitting. Warriors who can encourage the weak, drag the wounded, lift the tired. Warriors who know how to push through fatigue with truth instead of excuses. Warriors who don't need someone to yell at them to do the right thing — because they've built discipline into their bones.

God doesn't need your perfection. But He does want your preparation.

Because someday — and maybe soon — the enemy will test everything you claim to believe. And when that day comes, it won't matter how many sermons you've heard. It'll only matter how well you trained.

So if you've been coasting, wake up. If you've let your Bible collect dust, pick it up. If your prayer life has gone cold, fan it into flame. If you've relied on others to carry you, start building the strength to carry them.

You were not called to be passive. You were not saved to sit. You were rescued so you could become a rescuer. A warrior. A servant. A soldier of Christ.

So train. Not just for yourself. Train because the Body needs you. Train because the war is real. Train because your King is worthy. And train because when the call comes, when the mission lands, when the fire breaks out — the time for preparation will be over.

And you will either be ready…

Or you will be a risk.

CHAPTER 9

What is the Job of a Warrior?

FROM THE BEGINNING OF TIME, warriors have existed not merely as agents of destruction, but as defenders, protectors, and guardians of the highest ideals known to humanity — life, justice, freedom, and faith. The calling of a warrior is deeply biblical, profoundly spiritual, and eternally significant. A warrior, a soldier, and a fighter — each role echoes not only through the annals of human history but also through the divine narrative of scripture itself.

Warriors are not defined merely by the weapons they carry or the battles they fight. They are defined by their courage to stand between danger and innocence, between chaos and order, between evil and good. This is the biblical portrait of a warrior — one who fights because something worth defending exists.

Scripture repeatedly highlights that warriors and soldiers have a sacred role — not only on earthly battlefields but also in spiritual realms. And within God's great redemptive plan, the very heart of the Trinity — Father, Son, and Holy Spirit — operates with this warrior ethos at its core.

As a United States Marine and later as a military chaplain, I came to understand that this calling to be a warrior is not only consistent with Christian faith but essential to it. I have worn the uniform of a Marine. I have stood beside warriors. And I have fought battles — not always with weapons, but with truth, courage, and love.

This chapter will explore this timeless role, showing how God the Father, God the Son, and God the Holy Spirit embody these identities within the grand redemptive story. It will also show how this warrior ethos shaped my own life and ministry, particularly in my calling as a military chaplain — where I learned that sometimes the fiercest battles we fight are not with weapons, but with love, truth, courage, and compassion.

The Warrior Nature of God the Father — Defender of His People

Throughout scripture, God reveals Himself as a mighty warrior. Exodus 15:3 declares without hesitation:

"The LORD is a warrior; the LORD is his name."

This is not poetic exaggeration — it is divine self-disclosure. God is not passive in the face of evil. He does not stand idly by when His creation is threatened or when His people suffer. From Genesis to Revelation, God takes up arms against wickedness, not because He delights in war, but because He is committed to justice and love.

This is not poetic symbolism either; it is divine reality. From the plagues of Egypt to the collapse of Jericho's walls, God displayed His might to secure the freedom and safety of His covenant people. When Israel stood trapped between the Egyptian army and the Red Sea, it was not their strength that saved them — it was God's. In Exodus 14:14, Moses declared:

"The LORD will fight for you; you need only to be still."

Time and again, God took the role of military commander — leading, fighting, and securing victory for His people. In the wilderness, He fought against Amalek. In Canaan, He fought alongside Joshua. In the era of the Judges, He raised warriors like Gideon and Samson to defend Israel from its enemies.

This warrior identity of God is essential because it reveals His heart — He is not indifferent to the suffering of His people. He defends them. He fights for them. He protects what is holy.

In the Old Testament, God fought on behalf of Israel countless times. He sent hailstones from heaven (Joshua 10:11), confused enemy armies (Exodus 14:24), and empowered weak men like Gideon to conquer vast enemies with only 300 soldiers (Judges 7).

When Moses lifted his arms on the hill during Israel's battle with Amalek (Exodus 17:8-16), it was not human strength that won the day. It was divine intervention. Moses' raised hands symbolized dependence on God — the true warrior in every battle.

God the Son — Jesus Christ: The Perfect Soldier and Fighter

Many view Jesus as meek, humble, and gentle — and He was. Yet the fullness of His mission cannot be understood without recognizing that He was also a warrior. He came to fight the greatest enemies of humanity: sin, death, and Satan. Isaiah 42:13 paints this prophetic picture:

"The LORD will march out like a champion, like a warrior he will stir up his zeal; with a shout he will raise the battle cry and will triumph over his enemies."

Jesus is the fulfillment of this promise. His entire ministry was spiritual warfare — He cast out demons, confronted religious hypocrisy, challenged Satan in the wilderness, and ultimately conquered sin on the cross.

Jesus waged war from the moment He began His public ministry. He confronted demonic powers (Mark 5), challenged the hypocritical religious leaders of His day (Matthew 23), and ultimately fought the greatest battle on the cross. Colossians 2:15 declares the victory of Christ's warfare:

"And having disarmed the powers and authorities, he made a public spectacle of them, triumphing over them by the cross."

His resurrection was the ultimate military triumph — conquering death itself. In Revelation 19:11-16, Jesus is revealed as the rider on the white horse, Faithful and True, leading heaven's armies into the final battle against evil:

"With justice he judges and wages war... Out of his mouth comes a sharp sword with which to strike down the nations."

Here, Jesus is not the suffering servant but the victorious King, the divine warrior returning to claim what is His. Jesus returns not as a gentle teacher but as the conquering King, leading the armies of heaven. His robe is dipped in blood — a symbol of both His sacrifice and His victory.

The Holy Spirit — The Fighter Within Us

If God the Father fights for us, and God the Son fights for our salvation, the Holy Spirit fights within us. The Spirit equips, empowers, and strengthens believers for the daily battles of life. The Holy Spirit equips us with the armor of God — spiritual protection and weapons for the daily battles of life. Ephesians 6:10-18 outlines this armor:

- The Belt of Truth
- The Breastplate of Righteousness
- The Shield of Faith
- The Helmet of Salvation
- The Sword of the Spirit

These are not symbolic gestures — they are real spiritual tools that allow believers to stand against the attacks of the enemy.

Paul reminds us in Ephesians 6 that we do not wrestle against flesh and blood, but against spiritual forces of darkness. Yet, he also assures us in Romans 8:26 that: "The Spirit helps us in our weakness."

The Holy Spirit is our divine fighter, stirring up courage, perseverance, and faith within the hearts of every believer. It is through the Spirit that we put on the whole armor of God — the belt of truth, the breastplate of righteousness, the shield of faith, and the sword of the Spirit.

Every believer is a soldier in a spiritual battle — but we are not left unarmed or defenseless. The Spirit empowers us to resist temptation, to endure suffering, and to proclaim truth even in the face of opposition.

Warriors Fight to Defend, Protect, and Guard Ideals

The Bible is filled with men and women who embodied the warrior ethos — fighting not for themselves but for God's purposes. The biblical warrior is not defined by aggression but by their role as a defender of what is sacred.

- David fought not for glory, but to defend God's honor against Goliath's blasphemy (1 Samuel 17).

- Joshua fought to secure the promised land not for conquest alone, but to protect God's covenant people (Joshua 6).

- Even Peter, despite his misguided zeal, drew his sword in Gethsemane out of a heart to protect Jesus (John 18:10). Godly warriors fight to defend life, to protect the weak, and to guard what is holy.

These biblical warriors were not perfect — but they were courageous. They stood for God's honor and defended His people.

My Role as a Warrior, Soldier, and Fighter — A Chaplain's Calling

As a Marine and later as a military chaplain, I wear a uniform not to take life, but to defend it. I carry no weapons. My battles are fought in the hearts of the sailors and Marines I serve. I stand between darkness and light, between despair and hope, between loneliness and belonging. I recall vividly one such battle that forever shaped my understanding of this calling.

A Sailor in Distress — Stepping Into the Fight

It was late at night on a ship underway. The ocean was dark and quiet, but within the soul of one sailor, there was a storm raging louder than any thunder. I was called unexpectedly to the berthing area where this young sailor was found sitting alone, his hands shaking, tears streaming down his face. He was on the brink of taking his own life.

The voices of despair had convinced him that he was worthless, unloved, and forgotten. As I knelt beside him, I didn't come as a religious figure with fancy words. I came as a warrior — a defender of life, a protector of a soul in crisis. I looked him in the eyes and said: "You are not alone. You are not forgotten. There is a God who sees you. There is a chaplain here who will fight for you tonight."

We sat for hours — talking, praying, breaking through the lies that darkness had whispered to him. I became, in that moment, a soldier — not with weapons of war, but with words of life. That night, that sailor chose life. And that night, I understood with greater clarity than ever — I am a warrior, I am a soldier, I am a fighter — called not to destroy, but to defend.

The Eschatological Battle — The Final Victory

All of human history is moving toward one final battle — one last war where God will triumph over all evil. Revelation 20 tells us that Satan will be defeated forever. Christ will reign in perfect justice. The new heavens and new earth will be established — a kingdom where there will be no more death, mourning, crying, or pain.

Until that day, we are all in training. Life is our spiritual boot camp. The battlefield is all around us — in our homes, our workplaces, our relationships, and within our own hearts. But our Commander-in-Chief has promised: "To the one who is victorious, I will give the right to sit with me on my throne." (Revelation 3:21). We fight not for temporary victory — but for eternal glory.

Conclusion: The Warrior's Creed of the Believer

To be a Christian is to be a warrior. Not because we seek violence, but because we defend what matters most: life, truth, justice, and love.
— God the Father fights for us.
— God the Son fights for our salvation.— God the Holy Spirit fights within us.
— And we, as His soldiers, must fight for one another.

Whether in uniform or civilian clothes, whether on a battlefield or in a hospital, in a pulpit or on a ship — the calling is the same:
— Defend the weak.
— Protect the vulnerable.
— Guard the sacred.
— Fight the good fight of faith.
— For one day, the battle will be over.
— The trumpet will sound.

And the Commander of Heaven's Armies will say to us all: "Well done, good and faithful servant... enter into the joy of your Lord." (Matthew 25:23).

PART III

THE PREPARATION TASK

CHAPTER 10

Boot Camp is Not for Cowards

IT WAS NOVEMBER 2003 WHEN I STEPPED off the bus onto the famed grounds of Marine Corps Recruit Depot Parris Island, South Carolina. The night was heavy with a chill in the air, and the darkness seemed to press down on us with an intensity that matched the uncertainty in my heart. I was young. I was an immigrant, standing thousands of miles away from everything familiar.

I was searching for purpose, identity, and belonging in a country I was proud to serve. But nothing could have prepared me for the next thirteen weeks that lay ahead — a season that would shape me, break me, and rebuild me not only into a United States Marine but into a man who would later come to understand that every moment in boot camp was a mirror reflecting the Christian journey.

What I endured in that training was not simply a physical transformation; it was a spiritual revelation. Life, much like boot camp, is preparing us for something greater — for a final graduation day where all the suffering, hardship, and perseverance will be crowned with eternal joy.

The Yellow Footprints — Where The Old Life Ends

Every recruit's story begins the same way. The bus screeches to a halt in the dead of night. The silence is shattered by the thunderous voice of a Drill Instructor who wastes no time asserting his authority over the fragile minds of those who dared to enlist. "Get off my bus! Get on my yellow footprints!"

These were not kind suggestions; they were commands that demanded immediate obedience. I remember stepping down onto those iconic yellow footprints, and at that moment, something profound began to happen within me. The yellow footprints were not just paint on the pavement — they marked the line between who I was and who I was about to become.

In the Christian life, there comes a defining moment when we must step onto our own spiritual yellow footprints. It is the moment of surrender — when we realize that we cannot walk the journey of life in our strength, under our control, or according to our selfish desires. Jesus told His disciples in Luke 9:23, "If anyone desires to come after Me, let him deny himself, and take up his cross daily, and follow Me." Just like that moment on Parris Island, following Christ begins with surrender. It begins with dying to self. On the yellow footprints, our past does not follow us into the new mission. The old identity is stripped away to make room for something greater.

Receiving Phase — Breaking Down The Old Self

The days that followed our arrival were chaos, structured precisely to disorient us and to shatter any remnants of comfort. Civilian clothes were confiscated. Our heads were shaved. Our individuality was erased. We no longer referred to ourselves by our names but by "this recruit," stripping away personal identity until we were ready to receive the identity of a United States Marine. Everything that made us feel comfortable, safe, or secure was intentionally removed. I remember the mental battle that raged within me in those early days. It was humiliating, exhausting, and for many recruits, defeating.

But in that breaking down was a vital lesson. The same process happens in the life of every believer. When God calls us into His kingdom, He begins the holy work of sanctification — a process that breaks down our pride, our sin, and our fleshly desires. The Apostle Paul wrote in 2 Corinthians 5:17, "Therefore, if anyone is in Christ, he is a new creation; old things have passed away; behold, all things have become new." Becoming new requires the removal of the old.

God, like the seasoned Drill Instructor, knows that we cannot step into the new creation clinging to the remnants of our former selves. The breaking down is not cruelty; it is craftsmanship. God dismantles our false identities so that He might build us into vessels of honor.

Physical Training — Strength Through Struggle

Each morning at Parris Island began before the sun. We were awakened not by a gentle alarm clock but by the violent force of Drill Instructors storming through the squad bay. The physical training that awaited us was designed to break us down physically only to rebuild us stronger. Long runs in the freezing air, push-ups until our arms collapsed, navigating grueling obstacle courses — these were the daily rhythms of our lives. Every muscle ached. Every step felt heavier than the last.

And yet, day by day, strength began to emerge from struggle. Endurance replaced fatigue. Confidence replaced fear. The body that once shook under the simplest exercises now completed them with a pride that had been earned through sweat and pain. This is the paradox of the Christian life. Trials are not signs of abandonment; they are the tools of transformation. James 1:2-4 teaches us, "Count it all joy when you fall into various trials, knowing that the testing of your faith produces patience. But let patience have its perfect work, that you may be perfect and complete, lacking nothing."

God uses the resistance of life to build spiritual muscles. It is through hardship that we learn to trust, to depend on Him, and to become strong in the face of adversity.

Drill and Discipline — Walking In Step With Christ

Drill was one of the most critical aspects of Boot Camp training. We spent countless hours marching in perfect unison, responding instantly to commands, and learning to move as one unit. Every error was immediately corrected, every misstep addressed. The goal was simple — unity and discipline. A Marine does not move independently in formation; he moves as one with his brothers and sisters in arms.

The Christian life calls us to similar discipline. Galatians 5:25 says, "If we live in the Spirit, let us also walk in the Spirit." Following Christ is not about doing whatever feels right. It is about walking in step with the Spirit of God — being attentive to His voice, responsive to His correction, and moving in unity with the Body of Christ.

Drill taught us that small mistakes could have significant consequences. In spiritual life, discipline guards us against the dangers of sin. It aligns us with the mission of God and ensures that we move forward with purpose and precision.

Field Training — Surviving In The Wilderness

One of the most humbling and challenging parts of Boot Camp was field training. We left the barracks, our controlled environment, and entered the unforgiving world of the field. We slept in the dirt, ate rations, endured the elements, and learned survival skills. The comforts of even the harsh barracks were gone. We were forced to rely on our training, our fellow recruits, and our resilience.

God often does His most profound work in the wilderness. Moses spent forty years tending sheep in the desert before God called him to lead Israel. David hid in caves while fleeing Saul, being shaped by solitude and struggle. Even Jesus was led by the Spirit into the wilderness before His public ministry began.

Isaiah 40:31 encourages us, "But those who wait on the Lord shall renew their strength; they shall mount up with wings like eagles." The wilderness seasons of life are not punishment; they are preparation. They

CHAPTER 10: BOOT CAMP IS NOT FOR COWARDS

strip away distractions and force us to rely on the provision of God.

The Crucible — Enduring Until The End

The culmination of Boot Camp was The Crucible — a grueling 54-hour test that pushed us beyond exhaustion. We faced obstacle after obstacle, endured hunger and sleep deprivation, and navigated every test of endurance. My body felt broken. My mind teetered on the edge of quitting. But the presence of my fellow recruits, the drive instilled in me by my Drill Instructors, and the vision of what lay ahead kept me moving.

The Crucible reflects the spiritual race Paul describes in 2 Timothy 4:7: "I have fought the good fight, I have finished the race, I have kept the faith." Life itself is a crucible. It challenges every part of our being. It tests our faith, our perseverance, and our trust in God. But those who endure will be rewarded.

When we reached the top of The Reaper, the final hill of The Crucible, we were greeted not with shouts of anger but with quiet respect. Our Drill Instructors placed the Eagle, Globe, and Anchor in our hands — the emblem that signified our transformation into United States Marines. That moment was more than a victory; it was a new identity earned through perseverance.

Graduation Day — The Final Reward

Standing on the parade deck on graduation day was a moment of overwhelming pride and joy. My family stood in the crowd, their faces beaming with pride. The pain and struggle of the previous thirteen weeks faded in the light of that victorious day.

For the Christian, graduation day is coming. Revelation 21:4 promises, "And God will wipe away every tear from their eyes; there shall be no more death, nor sorrow, nor crying. There shall be no more pain, for the former things have passed away." Life is preparing us for a final graduation — the day we stand before Christ, free from sin, pain, and death, and enter into eternal rest.

Boot Camp in November 2003 shaped me deeply, not only as a Marine but as a man of faith. The journey of those thirteen weeks mirrors the journey of every believer. We surrender on the yellow footprints. We are broken down and rebuilt in Receiving Phase. We gain strength through struggle. We learn discipline and unity. We endure the wilderness. We press through the crucible. And one day, we will graduate.

This life is not the end. It is training for eternity.

On that final day, we will stand before Jesus Christ — our Commander and King — and receive the crown of life. And every trial, every hardship, every tear will be worth it when we hear the words that make every battle worth fighting: "Well done, good and faithful servant... Enter into the joy of your Lord."

Drill Instructor Sergeant Jesus

There are few figures in military culture more respected, feared, and ultimately loved than the Marine Corps Drill Instructor. They are the unrelenting force that shapes, molds, and transforms raw recruits — ordinary men and women from every corner of America — into the most elite fighting force in the world: United States Marines. But beneath the yelling, the stern faces, and the intense training schedules lies something much deeper. The Drill Instructor is not merely a taskmaster; he is a craftsman, a mentor, a guide, and a leader. His mission is not destruction — it is transformation.

When I look back at my time on Parris Island in November 2003, I can still hear the commanding voice of my Drill Instructors echoing through the humid, salty air. Their presence was inescapable. Their discipline was absolute. And yet, the older I become, and the more I reflect on my journey of faith, the more I begin to see a startling and humbling parallel — the Drill Instructor's role, in many ways, reflects the ministry of Jesus Christ. Christ too takes ordinary people — flawed, weak, unprepared — and through love, discipline, and guidance, transforms them into extraordinary instruments for the Kingdom of God.

Perhaps — just perhaps — the United States Marine Corps Drill

Instructor is one of the clearest earthly reflections of how Jesus works in the life of His disciples. What makes this even more remarkable is the reality that, in some cases, the military is accomplishing a level of transformation that many churches have failed to achieve. Jesus Himself may be using the military — a place often dismissed by religious circles — to accomplish His work of shaping leaders, instilling discipline, and creating men and women of honor.

Executing Training — Preparing for Battle

The first and most obvious role of the Marine Corps Drill Instructor is to execute the training programs designed to transform recruits into Marines. From the moment recruits arrive, they are immersed in a highly structured and demanding training environment. Every detail is intentional — every drill, every formation, every push-up, every lesson is building toward a specific goal: transformation.

Drill Instructors carry out this process with precision and dedication. They teach military drill to enforce unity and order. They push recruits to physical limits to build strength and endurance. They instruct in combat skills because a Marine must not only think like a warrior but act like one under pressure.

Jesus operates in the same way, though His battlefield is not merely earthly — it is spiritual. When He called the disciples, He called them not simply to believe but to train. He walked with them daily. He taught them through parables, confrontation, miracles, and even rebuke. The Gospels show us that Jesus was intentional — every lesson was preparing His disciples for spiritual warfare, leadership, and ultimately, sacrifice.

Matthew 28:19-20 records the Great Commission: "Go therefore and make disciples of all nations... teaching them to observe all things that I have commanded you." Teaching — executing spiritual training — is at the very heart of Christ's mission.

Where many churches have failed is in minimizing the role of disciplined training. The Marine Corps understands what many pastors forget — belief without discipline leads to weakness. Faith must be accompanied by preparation.

Instilling Discipline — Shaping Character

One of the earliest lessons every recruit learns is that discipline is non-negotiable. Drill Instructors demand absolute obedience. Not because they are tyrants, but because in the chaos of battle, hesitation or disobedience can cost lives. Discipline saves lives on the battlefield.

This discipline is not confined to physical behavior but extends to speech, thought patterns, respect for authority, and the ability to function under stress. Recruits are trained to follow lawful orders without question, trusting in the wisdom of their superiors and the mission at hand.

In the Christian life, discipline is equally vital. Jesus said in John 14:15, "If you love Me, keep My commandments." Obedience is not legalism; it is the expression of love and trust in God's wisdom. Hebrews 12:11 reminds us, "No discipline seems pleasant at the time, but painful. Later on, however, it produces a harvest of righteousness and peace."

The church has often struggled in recent decades to instill this kind of discipline. Messages of comfort, convenience, and emotionalism have replaced the hard, often painful work of spiritual formation. The Marine Corps does not cater to feelings — and neither did Jesus when calling His disciples to deny themselves, take up their crosses, and follow Him.

Perhaps the military succeeds where the church struggles because it understands that discipline is not unloving — it is essential.

Moral and Ethical Instruction — Honor, Courage, and Commitment

Integral to the Drill Instructor's role is teaching the Marine Corps' Core Values: Honor, Courage, and Commitment. These values are not optional slogans — they are the ethical foundation of every Marine's identity.

Honor teaches integrity — to do what is right even when no one is watching. Courage compels action in the face of fear. Commitment ensures perseverance through hardship.

These values mirror biblical teaching. Proverbs 11:3 says, "The integrity of the upright guides them." Joshua 1:9 commands, "Be strong and courageous." Galatians 6:9 urges, "Let us not grow weary while doing good."

While many churches preach these values, few environments enforce them like the Marine Corps. Drill Instructors engrave these virtues into recruits' hearts through repetition, practice, and accountability.

Christ calls believers to a higher standard — to reflect the ethical nature of God Himself. Yet the military often accomplishes this ethical instruction with far more consistency than many Christian institutions.

This reality challenges us to ask: Why is it that the Marine Corps — a secular organization — so often outpaces the church in instilling moral courage, integrity, and responsibility?

Serving As Role Models — The Living Example

Perhaps the most profound influence of the Drill Instructor is not what he says, but who he is. Drill Instructors are expected to be the embodiment of the Marine Corps' values. They are held to the highest standards of fitness, conduct, and appearance. Every recruit watches them. Every movement is a lesson. Every word carries weight because it comes from someone who lives what they teach.

Jesus was the ultimate role model. The Word became flesh and dwelt among us (John 1:14). He did not simply preach love — He lived it. He did not merely teach sacrifice — He embraced it on the cross.

In spiritual leadership, example is everything. Paul told the Corinthians, "Imitate me, just as I also imitate Christ" (1 Corinthians 11:1). The church suffers when leaders preach holiness but live in compromise.

Drill Instructors succeed because they lead by example. The church must recover this principle — that credibility comes not from position alone, but from integrity in action.

Developing Future Leaders — Preparing Warriors

The Drill Instructor's mission is not to create followers but leaders. Every recruit is trained not merely to survive but to lead in the future — to make decisions, to take responsibility, and to uphold the values of the Corps even when no one is watching.

Jesus spent three years developing His disciples into future leaders. He trained fishermen, tax collectors, and zealots to become apostles, evangelists, pastors, and martyrs.

Paul instructed Timothy in 2 Timothy 2:2, "And the things you have heard me say... entrust to reliable people who will also be qualified to teach others."

The military understands leadership development in ways that often escape the church. Churches sometimes focus on attendance rather than discipleship — on filling seats rather than filling lives with responsibility and calling.

Jesus built leaders.

So do Drill Instructors.

Mentoring And Guidance — The Role of Encouragement

While the image of the Drill Instructor is often harsh and intimidating, those who have endured Boot Camp know the secret beneath the surface — the Drill Instructor is also a mentor and guide. They know when to push and when to encourage. They have walked the path before. They understand the pain and doubt recruits face.

In the Christian life, Jesus is our ultimate mentor. He does not leave us alone in training. The Holy Spirit walks with us, encourages us, and reminds us of truth (John 14:26).

Effective churches are filled with spiritual mentors — seasoned believers who guide, correct, and encourage the next generation. The Marine Corps understands that transformation requires both challenge and care.

The Drill Instructor pushes hard because he knows what is required to win battles. Jesus disciplines us because He loves us (Hebrews 12:6).

The Military — Doing What The Church Cannot?

The sobering truth is that in many cases, the United States Marine Corps is accomplishing a level of transformation, discipline, leadership development, and ethical formation that many churches fail to achieve.

This is not a criticism to condemn the Body of Christ but a challenge to reflect deeply on our mission.

Could it be that Jesus is using the military — a secular institution — to accomplish spiritual formation in lives that the church has neglected or failed to reach?

Perhaps, as with the Roman Centurion in Matthew 8 — whose faith amazed even Jesus — God finds faith, obedience, and integrity in places we least expect.

Jesus called fishermen. Jesus called tax collectors. And perhaps today — Jesus is calling Marines. He is calling them not only to defend freedom but to reflect His character in the process. He is raising up leaders who have been forged in the fires of Parris Island, San Diego, or the battlefield — leaders who understand discipline, courage, commitment, and self-sacrifice.

Leaders who know what it means to endure to the end.

In many ways, Jesus is our Divine Drill Instructor. He sees in us what we do not see. He pushes us beyond comfort because He knows our potential. He disciplines us not out of cruelty but love. He trains us for battle because He knows the stakes.

And at the end of the journey — when life's Boot Camp is over — He will stand not as a harsh taskmaster but as a loving Savior.

He will place in our hands the crown of life. He will welcome us home. And we will know — every moment of struggle, every lesson of discipline, every tear shed in the training grounds of earth — was worth it. For we will stand not as mere recruits but as sons and daughters of the Living God — forever changed, forever loved, forever home.

CHAPTER 11

"Get Up and Walk"

IN THE SHADOWS OF BARRACKS, on the decks of Navy ships, and inside the minds of service members returning from deployment, there exists a spiritual crisis often unseen by the outside world. We think of soldiers as strong, stoic, and courageous, but beneath the surface of that hardened exterior is often a story of spiritual confusion, emotional brokenness, and moral paralysis. In my years serving in the military and as a chaplain, I have met countless young men and women who carry silent burdens — not just from war or combat, but from lifetimes of abandonment, neglect, fatherlessness, and spiritual starvation.

The military, in many ways, is a magnet for the marginalized. It draws in those who lacked opportunity, structure, or love. For many, enlisting is not simply a career decision — it is an escape. It's the first place they receive discipline, the first place they receive praise, the first place they feel part of a team. And while the Marine Corps or the Navy can teach a man how to fight, follow orders, and complete a mission, it cannot give him eternal purpose. That purpose comes from God. But when the church avoids these very spaces out of fear or misunderstanding, who is left to introduce that purpose?

Much like the pool of Bethesda in John 5, our modern military has become a collection of the spiritually sick — the blind, the lame, the paralyzed. Not physically, but spiritually. The church too often mirrors the Pharisees of that chapter, standing at a safe distance while issuing warnings: "Don't go there, you'll lose your faith." "The military is too secular." "It's not a place for a Christian." But Jesus never kept His distance from the sick. He walked straight into the colony of the broken.

John 5 and the Ministry of Fearlessness

The story of John 5 is far more than a miracle — it is a model. At the pool of Bethesda, we find a man who had been ill for thirty-eight years. That's nearly four decades of isolation, of being passed over, of watching others receive help while he remained stuck. "I have no one to help me into the pool," he tells Jesus. These words echo across the generations to our present moment, heard not just from the sick in first-century Jerusalem, but from soldiers in uniform today.

"I have no one to talk to."

"I've never had a spiritual mentor."

"My church back home stopped checking on me when I enlisted."

"I wanted to serve, but I was told it would ruin me spiritually."

And so, they sit — emotionally numb, morally disoriented, spiritually exhausted. They drown in busyness, in discipline, in deployments, but never in love. And when they look to the church, they see distance. When they reach for help, they are told they shouldn't have joined. And in this way, our spiritual Bethesda grows — a forgotten group in the church's mission strategy.

But Jesus did not fear Bethesda. He did not wait for the man to ask the right theological question. He did not test his motives. He simply asked, "Do you want to be made well?" Then He healed — even when the religious elite disapproved.

Ministry in the military is not for the faint of heart. It is not sanitized or quiet or predictable. But it is holy. And it is where Jesus still walks — if we are bold enough to follow Him there.

Seabee John — A Modern Example of Ministry in Bethesda

One of the most moving stories I've ever encountered in military ministry came from a Navy Seabee named John. He was a Seventh-day Adventist who deployed to the Philippines. There was no chaplain available. There were no Bible studies planned. There were no pre-approved programs, no digital screens, and certainly no stained-glass sanctuaries. But there was John — and there was a hunger for truth.

John, in his quiet but steadfast way, began gathering sailors each week. With few resources, often in scorching heat and with nothing but printed Scriptures, he led Bible studies for twenty service members. They sat on makeshift benches under tarps and cargo pallets, hungry for the Word of God. Some had never read a Bible before. Others hadn't prayed in years. But John was there — not because he was paid to be, not because anyone assigned him, but because God had placed a burden on his heart.

He did what many churches failed to do: he showed up. Week after week, John kept the group going, even when morale was low or the mission tempo was high. He wasn't a chaplain. He wasn't seminary-trained. But he was filled with the Spirit, and he answered the call.

This is the type of ministry Bethesda demands — fearless, patient, and deeply rooted in Christ. Seabee John did not avoid the spiritually crippled for fear of becoming one. He entered into their pain and invited them into healing.

The Church's Misguided Warnings

One of the great heartbreaks I've experienced is hearing from young Christians who wanted to serve their country — not out of rebellion or violence, but out of a desire to grow, to mature, to serve — was the condemnation of their parish. Instead of being blessed and commissioned by their churches, they were discouraged. "You'll backslide." "You'll be around sin." "You'll lose your way." While these concerns may come from well-meaning places, they often reveal a deeper flaw — the belief that God cannot work within the military.

But He can. And He does. If God could use Moses — raised in Pharaoh's palace, trained in Egyptian governance — to lead His people, then why not a Marine? If God could call Daniel to serve faithfully under multiple pagan kings, why not an Airman? If God could raise up Joseph to administer the affairs of Egypt, why not a Navy corpsman managing logistics?

The issue is not whether military service is dangerous. The Christian life is dangerous. The issue is whether we have the courage to believe God is already at work in unexpected places — and that He calls His people into those spaces not to be corrupted, but to be light.

Ministry Among the Spiritually Paralyzed

To minister in the military is to walk among the spiritually paralyzed. These are men and women who never had mentors, who came from fatherless homes, who saw addiction, abuse, and neglect long before they saw a Bible. They are hungry for discipline — but they're starving for grace. They can follow orders — but they've never been invited to follow Christ.

When I speak to fellow chaplains or believers in uniform, I often remind them: the person sleeping in the next bunk may be the next Peter, the next Lydia, the next Timothy. But without someone to guide, to witness, to disciple, they may never find their path. The Bethesda of today is filled with potential. The man who couldn't walk was healed — and stood up. And that is the mission: not simply to sympathize with the broken, but to help them rise.

Jesus as the Blueprint for Military Ministry

Jesus never feared reputation. He never tailored His mission to suit religious expectations. He walked into the unclean, the uncomfortable, the controversial. He touched lepers. He dined with tax collectors. He defended the woman caught in adultery. And in John 5, He stood in the middle of the rejected and restored a man everyone else had ignored.

For those who serve in the military and carry the name of Christ, your presence matters more than you know. You may be the only reflection of the Gospel someone ever sees. Your courage to pray, to speak, to invite others into your faith is more than admirable — it's missional.

And for pastors, elders, and church leaders: we must do better. We must stop fearing the military and begin supporting those who feel called to serve. Warn about the real spiritual risks, yes — but prepare them, equip them, bless them, and stand with them. Don't abandon them at Bethesda's gate.

Will You Enter the Colony of the Crippled?

Bethesda was not a holy place — it was a hurting place. But that is where Jesus chose to perform one of His most memorable miracles. He did not seek the temple courts that day. He sought the forgotten.
And He calls us to do the same. The military is not a spiritual wasteland — it is a mission field waiting for laborers. It is filled with the spiritually blind, lame, and paralyzed — not because they are weaker than civilians, but because they have often been left behind by the church.

But there is hope. There are still Seabee Johns. There are still young believers entering boot camp with Bibles tucked into their rucksacks. There are still chaplains who pray in silence over rooms full of sleeping soldiers. There is still healing at Bethesda — if we are willing to go.

The question isn't whether the military needs Christ. It's whether the church will have the courage to send Him there through His people.

The Crisis of Youth Disengagement and the Church

In recent decades, the church has witnessed a profound exodus of its young people. Barna Group reports that approximately 64% of U.S. adults aged 18–29, who were active in church as teens, have disengaged from church life altogether.[1] While 59% were already drifting a decade earlier, this increase signals an intensifying spiritual disillusionment.

To put this in perspective, nearly two-thirds of young Christians have effectively dropped out of the church after high school. Such attrition reflects more than a shift toward secularism—it signals a spiritual void growing deeper by the year.

When youth depart the church, consequences flow into personal and societal realms. The Centers for Disease Control and Prevention reports that 40% of high school students suffer from persistent feelings of sadness or hopelessness, and 20% seriously consider suicide, with 9% actually making an attempt.[2] These are not isolated statistics—they are the tragic outcomes of spiritual abandonment.

Barna's in-depth analysis identifies the underlying reasons for youth withdrawal: perceived lack of relevance, hypocrisy in leadership, absence of authentic relationships, institutional rigidity, and moralism disconnected from grace.[3] More than half of young adults, even those who still show up monthly, admit that church attendance is not essential to their faith.[4] These are not reversible trends; they are fundamental fractures.

Beyond the numbers lie individual journeys—stories of promising teenagers consumed by loneliness, anxiety, substance abuse, and broken identity. They needed belonging, purpose, and certainty—needs the current church model often fails to meet. So they leave—not necessarily

---

[1] David Kinnaman and Mark Matlock, *Faith for Exiles: 5 Ways for a New Generation to Follow Jesus in Digital Babylon* (Grand Rapids: Baker Books, 2019).

[2] Centers for Disease Control and Prevention, "Youth Risk Behavior Survey: 2023 Summary & Trends Report," accessed June 2025, https://www.cdc.gov/healthyyouth/data/yrbs/.

[3] Barna Group, "Six Reasons Young Christians Leave Church," 2011, https://www.barna.com/research/six-reasons-young-christians-leave-church/.

[4] Barna Group, "The Connected Generation," 2020, https://www.barna.com/the-connected-generation/.

because they reject faith, but because they cannot find it in their congregations.

Finding Purpose in the Military

Into this vacuum steps the military—a calling for structure, purpose, and meaningful contribution. For me, an immigrant seeking identity and direction, the Marine Corps Boot Camp of November 2003 was my turning point. I walked into that crucible with uncertainty, yet I walked out with a new sense of belonging, discipline, and contribution. It fulfilled my craving for purpose—while preparing me for broader forms of service.

Serving later as a Navy Chaplain, I discovered a powerful calling: the military not only refines character, but it can also be a stage for spiritual formation. On deck, in ships, and on deployment, I found young sailors craving approval, mentorship, and hope. I witnessed firsthand how those who were spiritually adrift found renewed identity as they engaged in small group Bible studies, baptism services, and one-on-one counseling. The military offered what youth ministries often fail to provide: a lived-out authenticity, shared suffering, and a mission that transcends personal gratification.

Today, as a Chaplain, I continue to find enormous joy and fulfillment. I've walked through grief with Marines during deployment, buried those who died in service, facilitated worship services in austere environments, and offered solace to those battling PTSD. This is the work that endures —souls marked by purpose, belonging, and God's presence even in adversity.

Why the Church Must Partner with the Military's Mission

The desert of youth disengagement demands a cross-cultural and missional approach. If we continue to treat military service as a threat to faith rather than a mission field, we will miss countless opportunities for transformation. Some specific ways churches can partner include:

— Commissioning young servicemen and women—praying for them publicly, sending them with blessing, and staying connected through phone calls and mentorship.

— Equipping believers for military ministry—training volunteers, providing resources, and forming small groups tailored to military realities.

— Honoring service spiritually and culturally—celebrating graduations, promotions, and deployment returns as spiritual milestones.

— Supporting returning veterans—offering re-integration classes, mental health care, spiritual renewal, and communal embrace.

In doing so, the church recognizes that the whole person—not just a Sunday version of them—is valuable to God. The military refines character, courage, and leadership; the church nourishes redemption, transformation, and eternal purpose.

Conclusion

When the church loses 64% of its youth, the casualty is not just doctrinal—it is personal, spiritual, and existential. These young people do not just leave churches—they fall into deeper challenges of despair and disaffection. Yet many find new life through military service—a calling unmasked as vocation, not menace.

My own story stands as testimony to this redemptive synergy—an immigrant who found direction and identity in the Corps, and purpose and sustaining joy in chaplaincy. Each day I see young sailors finding spiritual depth through authentic ministry in uniform, in trials, and in hope.

The crisis is urgent, but the solution is within reach—if the church will meet the military not with fear, but with grace and partnership, recognizing that God is already at work in this mission field.

CHAPTER 12

The Conversion Experience

IT WAS A QUIET NIGHT IN the countryside of Honduras —stars scattered across a vast black sky, the kind of celestial canvas that leaves a child awestruck. I was just ten years old, sitting on a cold cement bench, my legs too short to touch the ground, my heart heavy with longing. I hadn't seen my father in nearly a decade or my mother in almost eight years. The separation carved a void in me too deep for words.

That night, overwhelmed with emotion, I lifted my eyes toward the heavens and whispered a prayer so sincere that time seemed to stop around me: "Heavenly Father, if you let us reunite with our parents, I promise you I will serve you until I die."

I didn't fully understand the gravity of what I was saying, but something in my soul knew that moment mattered. What I didn't know was that even before I uttered that promise, God had already begun writing His response.

When the Promise Was Answered

A year later, what once seemed like a far-fetched dream became a miraculous reality. My parents were granted legal residency in the United States, and shortly afterward, my siblings and I received our green cards. After so many years of waiting, tears, and longing, we were finally going to be a family again. The moment of reunion was overwhelming.

I could barely recognize my father—his face etched with years of hard work and distance. My mother's embrace, though comforting, felt unfamiliar. And then there was a surprise: a younger sister I had never met before. Joy and confusion danced awkwardly together in our home. We were reunited, yes, but we were all different. Time had shaped each of us in ways the others could not yet understand.

As we tried to build a life together in America, I found myself slowly forgetting the promise I had made that night on the bench. In 1998, at sixteen years old, I stepped into a new world—one filled with overwhelming freedom, cultural disorientation, and social pressure. My parents, though well-intentioned and hardworking, lacked the cultural and financial knowledge needed to guide me through the transition.

Their own struggles—rooted in generational poverty and limited access to education—left them unable to offer the emotional and spiritual direction I craved. I don't blame them; they did the best they could. But I was lost. As the freedom of American adolescence wrapped itself around me, my connection with God unraveled. I began exploring, experimenting, and rebelling—not because I was angry at God, but because I had slowly stopped listening for His voice.

Forgetting the Promise

By November of 2003, I had accumulated debt, felt disillusioned with life, and was desperate for change. It was then that I decided to enlist in the United States Marine Corps. I didn't do it out of some noble sense of patriotism or divine calling. I did it because I needed direction, order, and financial security. Boot camp was brutal.

CHAPTER 12: THE CONVERSION EXPERIENCE

The thirteen weeks of physical punishment, emotional strain, and mental reprogramming tested every part of who I was. The yellow footprints I stepped onto at Parris Island became the foundation of a new identity. The drill instructors broke me down only to build me back up. I began to learn discipline, structure, resilience—and yet, I was still spiritually adrift.

I attended church services when I could, but nothing consistent. I tried to pray, but the words never came. I carried the uniform well, but underneath it, I was a man forgetting his vow to God.

And still, God pursued me.

He used people to reach me—fellow Marines who confided in me, trusted me, looked to me for leadership. I didn't understand it at the time, but God was preparing me. He was planting seeds through conversations, experiences, and responsibilities. I began to sense something shifting inside me. I still hadn't remembered the promise, but I started to feel its shadow growing longer across the path I walked.

An Awakening in the Midst of Success

Hoping for a fresh direction, I enrolled in college and chose architecture as my major. I loved the creativity, the logic, the structure of design. I even found success in the arts, becoming a professional dancer and performing at various venues. People admired me. I felt seen. But every standing ovation was followed by emptiness. The thrill faded quickly, and in its place came silence—the kind of silence that forces you to confront the truth of your soul. I was successful by many standards, but I was spiritually starving.

Then something changed.

I took an elective class called The Gospel of Jesus Christ. It seemed like a simple way to fulfill a general requirement. But what happened in that class would alter the trajectory of my life. One of our assignments required us to read The Four Faces of Jesus. I devoured that book. Each page felt like a divine appointment. I couldn't stop reading. It was as if my soul had been thirsty for years and now, finally, it had found water. I cried as I read, not from sorrow, but from recognition. And that's when it happened—I remembered.

I remembered the bench.

I remembered the sky.

I remembered the promise.

Seven years had passed since I whispered those words to God. I had broken my vow, but He hadn't broken His. In that moment of clarity, I realized just how patient and merciful God is. He had waited for me. He had orchestrated my life's journey—not to punish me for forgetting, but to prepare me to remember.

Theological Training Begins

I dropped my plans to become an architect and began studying theology. I enrolled at Griggs University and later continued my education at Antillean Adventist University in Puerto Rico. Each class, each late-night study session, and each deep dive into Scripture brought me closer to the God I had once promised to serve. I wasn't just learning theology —I was reclaiming my identity. I was finally becoming who I had vowed to be.

In many ways, my journey mirrored the life of Moses. Born under duress, raised in privilege, exiled by failure, and called back to purpose— Moses knew what it meant to forget a calling and be reminded by God. He ran from Egypt, buried in shame and fear, only to meet God in the solitude of the wilderness. I, too, ran—from Honduras to the U.S., from church to rebellion, from promise to performance. But like Moses, I encountered God again—not through a burning bush, but through burning conviction.

Called and Commissioned

The parallels continued to emerge. Moses tried to disqualify himself —"I stutter," he said. "I'm not enough." I told myself the same lies. I'm not educated enough. I've made too many mistakes. I'm too late. But just didn't need perfection; He needed surrender. And so, I gave Him what I had—my brokenness, my past, my potential. God accepted it all and used it as material for ministry.

Today, I serve as a parish pastor and Navy Chaplain—not because I'm better than anyone, but because God is faithful to everyone who says yes, even if that yes comes years late. I've counseled service members struggling with addiction, trauma, doubt, and despair. I've prayed with the broken, preached to the skeptical, and stood beside the dying. Every time I minister, I think back to that little boy in Honduras. I remember the cold bench, the wide sky, and the whispered promise. I see now that God never discarded that moment. He nurtured it, watered it, and watched over it until the day I was ready.

A Legacy of Grace and Calling

My story is a testament to what God can do with someone the world might overlook. I had nothing to offer but honesty. I was born into poverty, raised without parents, shaped by hardship, and scarred by rebellion. And yet, God used every part of my story as part of His plan. He does not choose based on résumé or reputation. He chooses based on readiness—and sometimes, readiness is born in the wilderness.

To anyone reading this who feels forgotten, disqualified, or directionless—know this: God remembers your promises, even when you forget them. He hears the prayers you pray as a child and holds them until you're ready to fulfill them as an adult. Your beginning may be humble, but your calling is holy. And the God who waited for me is waiting for you, too.

But I know how the story must look from the outside. "He joined the military and lost his faith." That's what many think. Some even say it with pity in their voice, as though I had fallen off the path and barely crawled my way back. I understand their reasoning. From the outside, my church attendance was inconsistent.

My choices weren't always Christlike. I spent some nights in local pubs with friends, and there were seasons where my spiritual fervor seemed all but extinguished. But assumptions have a way of mistaking silence for absence, and struggle for abandonment.

What they didn't see were the quiet Saturday mornings I found a church—sometimes small, sometimes unfamiliar—and slipped into the back pew to listen for a whisper from God. What they didn't know were the heartfelt conversations I had with local pastors who spoke into my life without judgment. What they missed were the moments I stood in front of a mirror in my dress blues, wondering who I was becoming—and if God still remembered the boy on that bench.

The truth is, I never stopped seeking God. I may have wandered in the wilderness, but I never let go of His hand. Even in my most distant moments, something inside me kept pulling me toward the light. I didn't walk away from faith; I was learning how to walk in it as a man, no longer as a child. My faith was no longer borrowed from my parents or my culture. It was becoming my own.

Faith in the Furnace

Military life, especially in the Marines, is a constant refining fire. It burns away pretense and reveals the core of who you are. The pressure, the discipline, the exhaustion—all of it strips you down to raw humanity. For me, it exposed my contradictions. I was a man who could lead Marines into formation but couldn't yet lead his own soul into consistent communion. I could command respect in uniform but struggled to speak the language of prayer. Yet God, like a patient potter, kept shaping me.

There's a beauty in realizing that spiritual formation isn't a single altar call or a climactic conversion moment. Sometimes, it's a series of gentle nudges, awkward prayers, and quiet realizations. Sometimes, it's sitting at a bar with a friend and realizing mid-conversation that your heart aches for something more. Sometimes, it's hearing a song in church that takes you back to the dirt roads of your childhood and the cement bench where you once made a promise.

God was present in all of it—in the joy and the guilt, the laughter and the longing. He didn't withdraw when I made mistakes. He drew closer.

He allowed me to taste the emptiness of chasing applause so that I would hunger for the richness of His presence. He let me dance on stages so I could discover that my soul was built for a different kind of audience.

The Unlikely Architect

When I declared architecture as my major, I thought I had found my calling. I enjoyed designing things. There was control in it, order, and purpose. But I was still designing buildings when God wanted me to build people. The structures I sketched on paper were nothing compared to the human souls I would one day counsel, guide, and lift up. I couldn't see it yet, but God could. And He patiently waited until I could, too.

The deeper I went into my coursework, the more disillusioned I became. Even my passion for dance and entertainment couldn't numb the quiet dissatisfaction growing in my heart. I was surrounded by noise, but my spirit felt muted. It was in that state that I took the class on the Gospel of Jesus Christ—not expecting transformation, just trying to fill a credit requirement.

But isn't that just like God? He meets us where we least expect Him. A routine elective became a divine encounter. A textbook became a prophetic voice. And suddenly, the long-lost voice of that ten-year-old boy returned to me. I remembered the bench. I remembered the stars. I remembered the promise.

Divine Memory

What astonishes me to this day is not just that I remembered—it's that God never forgot. Seven years had passed. Seven years of silence, distraction, compromise, and searching. And yet, He remembered. Not in a way that condemned me, but in a way that invited me. He remembered my vow, not to shame me for forgetting it, but to fulfill it with grace and purpose.

I sometimes imagine God watching over my journey—not with crossed arms, but with open hands. Hands that waited. Hands that covered me when I could've fallen farther. Hands that gently turned the pages of my life toward redemption. His memory is not like ours, soaked in bitterness and regret. His memory is redemptive. He remembers to restore, not to punish.

When I finally said yes again, I felt no scolding. Only welcome. Only warmth. Only the assurance that it's never too late to begin again. That is the essence of grace.

The Moses Mirror

I began to see my story in the life of Moses. Here was a man rescued as a baby, raised in privilege, driven by justice, but also marred by failure. Moses made a rash decision—killing an Egyptian—and fled into the wilderness. For forty years, he lived in obscurity, perhaps assuming his calling had expired. But God had not forgotten. He met Moses in the wilderness, not to remind him of his guilt, but to reignite his purpose.

I wasn't a murderer like Moses, but I had exiled myself in other ways—running from the discomfort of responsibility, hiding behind excuses, delaying obedience. And just like Moses, I argued with God when the call came again. "I'm not ready. I'm not holy enough. I've wasted too much time." But God silenced those voices with the same gentle command: "I will be with you."

That's all I needed.

I began my theological training, slowly at first, still unsure. But each class stirred something ancient within me. My textbooks became scripture. My professors became shepherds. My old doubts became new questions that led to deeper faith. I wasn't just learning theology—I was rediscovering identity.

A Chaplain's Calling

Eventually, I accepted my dual call—to serve as a parish pastor and as a Navy Chaplain. And now, when I walk the corridors of ships or sit across from service members in pain, I recognize the same lostness I once felt. I see in them the spiritual hunger I once tried to ignore. And I get to offer something real—not religion for religion's sake, but relationship. Not empty rituals, but meaningful encounters with the living God.

I've held sailors in prayer after they lost a parent. I've stayed up late counseling Marines struggling with depression. I've stood in the hot sun officiating weddings and in the cold of the night delivering eulogies. And in every moment, I know that this is what God was preparing me for when He waited for me to remember the promise.

Every benediction I give, every sermon I preach, every silent prayer I whisper over someone who feels unseen—it all traces back to a starlit night in Honduras when a ten-year-old boy made a promise that God never forgot.

Purpose in the Wilderness

Sometimes, people think that their mistakes disqualify them. That the time they've lost can't be redeemed. That the promise they broke disqualifies them from service. But that's not how God works. He specializes in resurrection—not just of bodies, but of dreams, of callings, of destinies.

My journey wasn't clean. It wasn't linear. But it was holy. And yours can be too.

I want to speak directly to someone reading this who thinks, "It's too late for me." It's not. If God waited for me, He will wait for you. If He redirected Moses at a burning bush, He can meet you in your living room, your barracks, your car. If He used my failures to mold me into a minister, He can use yours to birth something beautiful.

You don't have to be perfect to be called. You just have to be willing.

God's Faithfulness, Not Mine

At the heart of this chapter is not my discipline, my loyalty, or my eventual obedience. At the heart of this story is God's faithfulness. It's His willingness to wait. His commitment to pursue. His refusal to let go, even when I did. It's His relentless grace that reached through my rebellion, my confusion, and even my pride to remind me that He finishes what He starts.

So when I wear my uniform today—whether it's clerical or military—it's not a badge of my achievements. It's a symbol of His patience. His mercy. His vision. I am not a pastor because I earned it. I am a pastor because God kept His promise.

And He still does.

PART IV

GOING TO WAR

CHAPTER 13

The Morse Code

BEFORE I UNDERSTOOD IT, Morse code sounded like chaos — a random sequence of clicks, beeps, or flashes without form or meaning. It felt disconnected, like broken language. The first time I heard it, I was watching a war documentary in high school. A naval operator was hammering away on a metal key, sending dots and dashes through the static. I remember thinking, How could anyone make sense of that?

And yet, across oceans, someone on the other side was receiving those exact taps — interpreting them instantly into words, directions, and commands. That scattered noise was, in reality, an intimate message. A lifeline. It wasn't just communication; it was survival. And during war, it often spelled the difference between victory and defeat.

Morse code was designed with beautiful, brutal simplicity: short and long signals representing letters. Alone, a dot meant nothing. A dash meant nothing. But together, they were power. Together, they carried clarity — to those who were trained to understand it.

Divine Code

In World War II, these signals shaped battle outcomes. Morse code was used aboard naval ships, in aircraft, and on frontlines where the spoken word was too dangerous or impossible. At the Battle of Midway, critical intelligence was transmitted via Morse code and cipher systems. The U.S. Navy, having broken parts of the Japanese code, used subtle signals to deceive and trap enemy carriers, leading to a decisive turning point in the Pacific Theater. The messages weren't long. They weren't shouted. They were short bursts of light and sound — coded speech that could only be understood by those who had studied the pattern.

The power of Morse wasn't in how loud it was, but in how precise. It wasn't meant to be understood by everyone. In fact, its purpose was to hide the message from enemies while delivering life-saving instruction to the trained. And it worked. Countless lives were saved, countless missions completed, because someone knew how to read a code the enemy couldn't break.

The more I learned about this — not in documentaries, but in real-life training and history — the more I began to see something startling in my walk with God: He speaks in code, too.

Not everyone likes to admit that. We want to believe God always speaks clearly, loudly, and plainly. But He doesn't. At least not always. More often than not, He speaks in what seems like chaos — in hardship, in delay, in unexpected placement. He speaks through trials. Through pain. Through what looks like contradiction.

And at first, it all feels like random noise.

When I first joined the military, I couldn't make sense of what God was doing. Why would a God of peace call someone into an institution of war? Why would He place someone who hated violence into a system built on combat readiness and destruction? Why would He allow someone to be broken in order to be built?

It didn't make sense.

And then one day — in the middle of field training — I felt it. Not an answer. Not a moment of clarity. But a pattern.

CHAPTER 13: THE MORSE CODE

I was learning obedience. Discipline. Endurance. Brotherhood. Humility. Things I had prayed for — but had expected to learn in pews and prayer meetings, not through rifles and rucksacks. It dawned on me that God hadn't gone silent — I just hadn't been trained to recognize His voice in this format.

God wasn't shouting. He was tapping. And each moment, each day, each hard thing became a signal in the code.

What I once saw as chaos, I now saw as divine communication.

Romans 11:33 says, "Oh, the depth of the riches both of the wisdom and knowledge of God! How unsearchable His judgments and His paths beyond tracing out!" That verse always sounded poetic until I lived it. Until I felt like I was walking a path that made no earthly sense — only to discover God was guiding me by a map written in divine Morse code.

See, when you're on the receiving end of God's voice, and it doesn't match your expectations, your first instinct is to doubt it. You want burning bushes, not battlefield barracks. You want angels and visions, not field ops and firewatch. But God doesn't owe us clear speech. He gives us coded faith because it sharpens our hearing.

Morse code is hard to decipher when you haven't trained for it. So is God's voice when all you've ever wanted is convenience.
I wonder how many people have walked away from the very calling God gave them because the code didn't make sense at first. I wonder how many believers are asking, "Why am I in this season? Why this pain? Why this environment?" without realizing they're inside the very message God has been trying to send.

Maybe it's not chaos. Maybe it's code.

I've come to believe that what we often interpret as demonic, profane, or irrational might actually be the platform of divine communication. Because God doesn't just speak through miracles — He speaks through mystery.

When I look back now at my enlistment, at the early confusion, the moments of hardship, the things I didn't understand — I see the message. Not all of it, but enough to know it wasn't random. My discomfort was part of a pattern. My hardship was part of heaven's syntax. God was using a language I hadn't yet learned to listen for.

And maybe that's the question every Christian in uniform should wrestle with: What if your service — with all its contradiction and difficulty — is God's way of speaking not just to you, but through you? What if your obedience to join the military is His code to reach someone else? What if your station is a sentence in someone else's salvation?

We say things like, "God works in mysterious ways," but rarely do we stop to interpret the mystery. We want the message, but not the method. Yet God has always used strange vehicles to deliver truth: a donkey, a prostitute, a Roman cross, a persecutor named Saul. He speaks through what seems offensive. He delivers through what seems unqualified.

So maybe the military — with its violence and profanity, with its loss and machinery, with its order and chaos — is exactly the kind of place God speaks from. Not because He delights in war, but because He knows how to encode glory in the middle of what looks like madness.

Maybe divine Morse code isn't about dots and dashes — maybe it's about meaning wrapped in mystery. Maybe the flashes of suffering, the silences between assignments, the moments of obedience in obscurity are part of the sentence He's still forming.

And maybe the reason Satan can't stop it is the same reason Japanese forces couldn't crack our codes — because they weren't written for them. Because only those trained to hear will ever understand.

The Confusion of the Code

If Morse code taught me anything, it's that clarity is not a prerequisite for effectiveness. In fact, its brilliance lies in its confusion. What looks like nonsense to the enemy becomes life to the one who knows the key. And that principle has won wars.

During the Second World War, secrecy in communication was as valuable as ammunition. Armies needed a way to transmit orders without revealing their intent. The Allies mastered this with a combination of ciphers, codebooks, and radio silence, but the most unbreakable system came from a source no one expected: the Navajo Code Talkers. The U.S.

CHAPTER 13: THE MORSE CODE

Marines recruited Navajo speakers to develop an entirely new code based on their native language.

It was so unique, so unfamiliar to Japanese cryptographers, that it became virtually impossible to crack. Messages about troop movements, artillery positions, and secret operations could be sent in seconds with absolute confidence. Major General Howard Connor, a signal officer during the Battle of Iwo Jima, later said, "Were it not for the Navajos, the Marines would never have taken Iwo Jima."

The genius of that code wasn't that it was complicated — it was that it was alien to the enemy. They could intercept every syllable, but without the key, it was meaningless noise. Victory hinged on the difference between hearing and understanding.

That reality is exactly how God's communication feels sometimes. People who don't know Him can hear the same sermons, read the same verses, even experience the same events, but it's noise without revelation. The code is audible but unintelligible until you've been trained by the Spirit to recognize its patterns.

Isaiah 55:8–9 captures this perfectly: "For my thoughts are not your thoughts, neither are your ways my ways, declares the Lord. As the heavens are higher than the earth, so are my ways higher than your ways and my thoughts than your thoughts." God isn't just smarter. He's operating in a dimension we don't naturally understand. His messages arrive wrapped in symbolism, timing, and paradox, just like prophecy.

Take Daniel's visions or John's Revelation. Beasts with horns. Numbers with hidden meaning. Angels measuring cities with golden rods. For centuries, those messages were mocked as nonsense. Even today, many read them and only see confusion. Yet to those trained in Scripture, in humility, in prayer, patterns emerge. Prophecy is like divine cryptography — dots and dashes of imagery pointing to a larger message that only becomes clear when God illuminates it.

I've learned that God often works like a code talker in our lives. He uses experiences, environments, even people we'd never expect as His vocabulary. To an outsider, it looks like chaos. To the one who knows His voice, it becomes a roadmap. When I first joined the military, the idea that

God could use something so violent and imperfect to shape something holy seemed impossible. But looking back, I see how each assignment, each test, each moment of discipline formed a word in His message to me.

It's the same with prophecy. When God gave Daniel visions of empires rising and falling, or when Revelation painted cryptic pictures of beasts and plagues, He wasn't trying to scare or confuse His people. He was encoding a warning and a hope in a form the enemy could not destroy. The Amazing Facts ministry often compares biblical prophecy to a coded letter — symbols that might seem strange at first but reveal a stunningly consistent message when interpreted correctly. Prophecy wasn't random. It was strategic. Like a general's battle plan, hidden from enemy eyes but broadcast clearly to the troops who had the key.

That's how divine Morse code works. The world hears static. The enemy intercepts signals. But the Spirit translates.

I remember sitting in a chapel service during training, exhausted and wondering why I'd even come. The chaplain read from Romans 8:28, a verse I'd heard a thousand times: "And we know that in all things God works for the good of those who love him, who have been called according to his purpose." But that day, in my fatigues, with dirt under my nails and sweat on my back, it hit different. It wasn't a cliché. It was a code. All things. Even this. Even here. Even the orders I didn't like, the pain I didn't understand, the isolation I hadn't chosen. All of it was part of the pattern.

The enemy may hear your prayers. He may see your movements. He may know your weaknesses. But he can't decode God's intent. He can't unravel the purpose woven into your placement. Just as Japanese cryptographers sat in rooms full of intercepted messages, scratching their heads in frustration, so too does hell watch God's people move into places and positions that make no sense on paper — only to realize too late that those very moves were setting up a breakthrough.

That realization reframed my entire military journey. What I once saw as detours I began to see as dots and dashes. What I once called random, I began to call a message. The code was working — not because I understood every part, but because I trusted the Sender.

And that's the invitation for every believer, especially those in uniform: don't despise the code. Don't reject the mystery because it feels confusing. Train your ear. Lean in. God's communication is not always linear, but it is always intentional. What seems unreasonable may be His strategy. What seems profane may be His instrument. What seems like war may be His way of bringing peace to someone who would never have heard the gospel otherwise.

Morse code in war was simple: dots and dashes. Yet it defeated empires because its key was hidden. God's Morse code may look like pain and paradox, but its key is hidden in Christ. If you stay close to Him, you'll start to hear the words behind the noise.

Messages in the Fire

The first time I truly heard God's voice, it didn't sound like anything I'd imagined. It wasn't a gentle whisper, or a powerful shout from heaven. It came in the form of fire — confusion, hardship, and silence. But somehow, in the middle of it, something in my spirit knew: this was Him. I hadn't misstepped. I was exactly where I was meant to be. It just didn't feel like it.

We don't often associate God's voice with fire anymore. We expect peace. Clarity. A clean, polished confirmation. But in the Scriptures, God speaks through flames far more often than through gentle winds. Think of Moses — it was in the burning bush that God revealed His name and mission. Or Shadrach, Meshach, and Abednego — thrown into a literal furnace, yet it was in that heat that God appeared.

Pain is often the punctuation in the divine code.

As I continued my time in service, I came to realize that some of God's clearest messages came when I was at my lowest — during deployments, personal losses, separation from family, and even through watching others suffer. The heat of those moments burned away my assumptions. It stripped the surface-level understanding I had of God. It left behind only what was real — what was forged.

During WWII's D-Day invasion, communication was everything. Entire battalions waited for brief, coded transmissions to move. Lives depended on receiving the right signal at the right time. Soldiers crouched behind bunkers, wet and trembling, clutching their weapons, waiting for one tap of the key — a burst of sound, a flash of light — before launching into history. They didn't have the whole battle plan. They just had the next move. And that move would change everything.

That's how it often feels walking with God through fire. You don't get the full map. You just get a signal — a spiritual "dot" or "dash" in the form of an unexpected phone call, a verse you weren't looking for, or a door that slams shut. And if you haven't trained yourself to listen through the crackling noise of fear or exhaustion, you might miss it. But when you catch it — when the pattern becomes clear — you realize you've been moving inside a message all along.

That's how it often feels walking with God through fire. You don't get the full map. You just get a signal — a spiritual "dot" or "dash" in the form of an unexpected phone call, a verse you weren't looking for, or a door that slams shut. And if you haven't trained yourself to listen through the crackling noise of fear or exhaustion, you might miss it. But when you catch it — when the pattern becomes clear — you realize you've been moving inside a message all along.

In those moments, I remembered the words from Romans 8:28: "And we know that in all things God works for the good of those who love him." It's a comforting verse, but one often stripped of its battlefield roots. Paul didn't write that from a vacation. He wrote it from a life wrecked by suffering — shipwrecks, floggings, betrayal, imprisonment. Yet in all those hardships, Paul decoded a greater message: God works through the fire. He doesn't bypass it.

The fire was not a failure in the system. It was the system. It was how God signaled something deeper. Something eternal.

The military gave me a lens to see this clearly. We were trained to suffer — to keep moving when sleep-deprived, to complete the mission with aching backs, to trust the order even when we couldn't see the objective. It was frustrating. It felt unfair. But every one of those moments was forging the kind of obedience that would hold when everything else collapsed.

And isn't that the Christian life?

When you've prayed for purpose, and you're handed pressure instead. When you ask for peace, and He sends you into a fight. When your prayers seem unanswered, and yet every door is moving you somewhere. That's not God ignoring you — it's Him communicating on a frequency most people have stopped tuning into.

That's divine Morse code.

Think of Joseph in the Old Testament. God gave him dreams as a teenager. But instead of a straight path to the palace, Joseph was thrown in a pit, sold into slavery, falsely accused, and imprisoned. If that was you, what would you hear in those moments? Silence? Abandonment? Or would you, like Joseph, learn to trust the pattern of God's hand even when the meaning seemed lost?

By the end of his life, Joseph saw the message behind the mystery. He looked into the eyes of the very brothers who betrayed him and said in Genesis 50:20, "What you meant for evil, God meant for good." That was Joseph's divine code decoded — years later, through fire and silence, the signal became clear. It was all part of the message.

And here's where it becomes deeply personal: what if the pain you're walking through isn't punishment — it's a paragraph in someone else's deliverance? What if the fire you're enduring isn't random — it's a refinement? What if your enlistment, your obedience to step into the battlefield of military life, is not a detour from God's glory but a deliberate part of it?

The world calls it trauma. God might be calling it testimony. The world sees violence. God sees vessels being shaped.

In Revelation, John sees a scroll sealed with seven seals — a message no one could read. The elders wept because it seemed like no one could open it. But then, a voice announced, "The Lion of the tribe of Judah has triumphed." And only the Lamb could open the scroll. Only Jesus could break the seal.

Some messages in your life will remain sealed until Christ Himself breaks them open. Some seasons will make no sense until He gives revelation. But just because you can't read the code now, doesn't mean you're not part of it.

I've learned not to fear the fire. Because the fire is where the clearest signals are sent. And I've learned not to curse the code — because what looks like confusion is often salvation in disguise. A delayed letter. A denied order. A failed relationship. All of it can be God's dots and dashes spelling out a message only seen in hindsight.

We think God only speaks in comfort. But He speaks in command lines. In wounds. In the weary miles we march without explanation.

And if you listen, really listen — you'll start to hear the pattern.

The Enemy Cannot Decode Obedience

If there's one thing the battlefield teaches you, it's that obedience is not about understanding — it's about trust. Orders don't come with explanations. You don't get a briefing on every outcome before you move. You just move. You act. You follow. You do what's been asked of you because the one who gave the order knows something you don't.

That's what makes spiritual obedience such a powerful weapon — and such an impossible code for the enemy to crack.

In war, messages are intercepted all the time. The enemy picks up radio chatter, watches movements, taps into lines. But the messages that work — the ones that win wars — are those encrypted so deeply that even when they're heard, they can't be understood.

During World War II, this truth was brutally tested. At one point, the Japanese navy intercepted U.S. transmissions and wrongly believed the next target would be Midway. But they weren't sure. So to confirm it, the Americans sent a fake Morse-coded message from Midway that their water supply system had failed. Shortly after, Japanese communications repeated that "AF was short on water" — confirming to U.S. analysts that "AF" was indeed Midway. That one decoded lie allowed the U.S. to prepare an ambush, turning the tide of the entire war in the Pacific.

The enemy had access. They were listening. But they misread the

CHAPTER 13: THE MORSE CODE

message — and it cost them everything.

I think about that story a lot when I reflect on the enemy of our souls. Satan hears our prayers. He sees our worship. He watches our routines. But what he can't decipher — what he has never been able to anticipate — is raw, unscripted obedience.

When God tells you to take a step that doesn't make sense — to forgive someone who doesn't deserve it, to speak when silence would be easier, to enlist when you planned to go into ministry, to stay when every cell in your body wants to run — and you obey anyway, you activate a frequency the enemy can't trace.

Satan traffics in logic, ego, and impulse. But obedience — faith-based, costly, unreasonable obedience — throws off his entire playbook.

Think of Abraham. God tells him to leave his homeland, his family, and go "to a place I will show you." Not "a place I've shown you." Not "a place you'll love." Just… go. Abraham obeys. And that obedience becomes the launchpad for a nation, a covenant, and a redemption plan that would stretch through generations. Satan couldn't stop it — because he couldn't see it coming.

Then there's Esther. A Jewish girl in a Persian palace. No one knew who she truly was. And when Haman plotted genocide, God's coded message came through her uncle: "Perhaps you were born for such a time as this." What does she do? She obeys. She risks death. She walks into the king's presence unsummoned. Her courage, wrapped in obedience, rewrites history.

And of course, there's Jesus — the ultimate divine code in human flesh. Born in a manger, not a throne. Riding a donkey, not a warhorse. Crucified, not crowned. To the religious leaders of the day, He didn't look like a king. To Rome, He didn't look like a threat. To Satan, He looked like a vulnerable man in a weak position. But what none of them realized was that every step of Christ's obedience was part of a coded mission — a pattern so divine that even death couldn't decode it.

The cross looked like defeat. It was actually the detonation of salvation.

That's what obedience does. It confuses hell. It ruins demonic forecasts. It disrupts expectations.

Obedience isn't glamorous. It's not always applauded. It rarely makes sense in the moment. But it is the single most prophetic thing a believer can do. And it's the one thing Satan cannot counterfeit.

When I joined the military, I wrestled with the decision. I had people in my life who couldn't understand it — Christians who said, "How can you follow Christ and carry a weapon?" They didn't understand that I wasn't following a cultural pattern. I wasn't seeking identity or power. I was responding to a signal that wasn't meant for them. It was obedience — and it was encoded.

To this day, I may not understand all the reasons why I was called into service. But I know that in my time in uniform, I've been places where the gospel needed to be spoken. I've prayed over bodies broken by war. I've counseled hearts that pastors would never have reached. I've seen darkness up close — and I've seen light shine brighter because of it.

The enemy never saw that coming. Because he never sees obedience coming.

He expects you to act like him — prideful, reactive, self-centered. He expects you to preserve yourself. To question the call. To delay the mission until it fits your comfort. But when you obey despite confusion — when you say "yes" before knowing "why" — you're walking in a language he can't decode.

That's the divine brilliance of it all.

In Jeremiah 29:11, God says, "For I know the plans I have for you…" He doesn't say you know them. He doesn't lay them out in full. He simply asks for trust. Faith. Obedience.

In the military, we say, "Ours is not to question why; ours is but to do and die." That's not blind loyalty — that's functional submission. Because we trust that someone sees more than we do.

Spiritually, the same holds true. I don't follow God because I understand everything. I follow Him because He is the Commanding Officer of eternity — and His orders always lead to victory, even if they pass through a battlefield.

So when you feel the tension between what God is asking and what you expected, don't try to decode the whole plan. Just act on the next command. Just follow the last signal. Just trust the tone in His voice. Your obedience may be the very key that unlocks a revival, a rescue, or a revelation in someone else's life.

And best of all? The enemy still won't see it coming.

When Heaven Sends a Signal

The genius of Morse code wasn't in its volume — it was in its precision. A message sent over noisy battlefields, through storms, or across radio static didn't need to be long or loud. It just needed to be accurate. The receiving end didn't need a speech, just a signal. And if the receiver was trained, even a single dot or dash could carry mission-critical meaning.

That's how God speaks.

Over time, I've learned that heaven doesn't need to raise its voice to be heard. It only needs trained ears. God's signals aren't always wrapped in miraculous moments or obvious signs. Often, they're subtle. Quiet. Repetitive, even. A word from a friend. A Bible verse repeated in passing. A heaviness that won't lift until you obey. A delay you didn't want. An assignment you never asked for. Alone, those moments might feel meaningless. But together, they spell out a divine message only those attuned to heaven's frequency can decode.

Jesus said it this way in Revelation 3:22: "He who has an ear, let him hear what the Spirit says to the churches." Not just anyone will hear. Only those who've trained their ears. Who've learned to distinguish spiritual signal from emotional noise. Who've disciplined themselves to recognize the tone of their Shepherd's voice, even when it's disguised in silence or suffering.

The more I've walked with God, the more I've realized that His signals don't always explain — they command. And obedience becomes the confirmation.

I think about those who fought in World War II, crouched in foxholes, listening to single-letter transmissions tapped out by trembling fingers.

They didn't need explanations. They just needed the next move. And when the move came — they trusted it. They acted on it. Not because they understood the entire plan, but because they believed in the one sending it.

That's what faith is. Faith comes by hearing — not by having the whole picture. We don't walk by clarity. We walk by code. And sometimes, the greatest sign that heaven is speaking is that we feel the tension to respond before we understand what we're responding to.

Maybe you've been there. Maybe you are there now. Standing at the edge of something you didn't plan for. Carrying a weight that doesn't make sense. Hearing a faint signal in your spirit but questioning if it's really God. Let me say this clearly:

He is speaking.

He may not be shouting. He may not be using the method you expected. But He's speaking. In the delay. In the discipline. In the closed doors. In the unexpected orders. In the painful reroutes. In the uncomfortable promotions. In the early morning nudges that won't let you go.

Sometimes, all He sends is a single character — a "dot" in your journey. A prompt. A check in your spirit. A moment that doesn't make sense yet. But that signal, if obeyed, may carry the full weight of His will.

In my own life, I can look back now and see the dots and dashes strung across the years — a Divine Morse code that carried me from confusion to clarity, from aimlessness to assignment. I didn't always understand the signal at the time. But looking back, I can see the message now.

God used the military — an institution of discipline, obedience, hardship, and sacrifice — to train my ears. To tune my heart. To prepare me not just for war, but for witness. I've prayed with soldiers who'd never set foot in a church. I've carried the gospel into places where traditional ministry would never reach. And it all started with a whisper I almost ignored.

A signal.

And what's more — the enemy still doesn't understand it. He sees the same movements. Hears the same words. But he doesn't know the Sender. He can't decipher the obedience of a yielded heart. And because of that, every moment you choose to respond to God's prompt — no matter how small — becomes a declaration of war against hell's strategy.

So listen.

Train yourself to hear heaven's frequency. Get quiet enough to detect the tap of eternity against your heart. Be still enough to hear the whisper behind the noise. Stop waiting for the message to come how you expect it — and start looking for it in how it already is.

A flat tire might be a delay that saves you.

A sudden reassignment might be a placement for someone else's miracle.

A closed door might be a divine protection you won't understand until years later.

A long season of silence might be heaven tapping out a message you're only now ready to hear.

To those untrained, it's noise. To those who listen, it's the voice of God. So the next time you wonder where He is, the next time you question if this detour is punishment or purpose, remember this: Morse code saved lives not because it was complex, but because someone was trained to interpret it. You don't need to know everything. You just need to hear clearly.

Because what sounds like chaos to hell… might be a battle hymn to heaven.

CHAPTER 14

Combat Strategies

THE FIRST THING THEY DRILL into you at the U.S. Navy War College is this: no war is won by strength alone. Victory belongs not to the strongest army, but to the smartest commander — the one who sees the battlefield before a shot is fired, who calculates not just firepower, but timing, terrain, deception, and morale.

Strategy wins wars.

And strategy saves lives.

Every Battle Begins with a Plan

I still remember sitting in that room, the walls lined with decades of combat doctrine, instructors parsing through battles long finished but never forgotten. We studied Clausewitz, Sun Tzu, naval doctrine, multi-domain operations, and the anatomy of failed and successful campaigns. One point rose again and again like a lighthouse in stormy waters: whoever has the clearest plan — and sticks to it — wins.

One of our case studies was the Battle of Midway — the Pacific turning point of World War II. On paper, Japan should have won. They had more ships, more experience, and had struck fear across the Pacific. But the U.S. had what they didn't: a strategy grounded in intelligence, timing, and misdirection. Naval cryptographers had intercepted and decoded Japanese signals. They knew Japan was targeting Midway. The U.S. Navy set a trap.

The result? Japan lost four carriers in a single day. It was the beginning of the end for their naval dominance. Not because we had more power — but because we had better insight and the discipline to trust the plan.

That concept haunted me as I lay in my bunk one night, flipping through Daniel 2 and Revelation 12. The parallels gripped me.

Here was God — not reacting to sin, but outmaneuvering it. Not improvising salvation, but orchestrating it with precise, prophetic foresight. The war against sin wasn't won at Calvary by accident. It wasn't Jesus making the best out of a bad situation. It was a long-game strategy executed with flawless timing, born in eternity and prophesied throughout Scripture.

And suddenly, everything we studied in warfare clicked into place with the gospel.

God is not just a Redeemer — He is a Strategist.

In Christian eschatology, the plan of redemption is not a loose patchwork of moral teachings. It is a combat strategy — a cosmic operation to vindicate God's character, liberate creation, and eliminate evil without violating free will.

That is far more complex than it sounds. Anyone can eliminate evil with force. But what God is doing? He's winning the war by proving, with divine patience and justice, that His government is built on love — and that love is the only sustainable power in the universe.

It began with rebellion. Lucifer, heaven's highest angel, initiated a coup in God's command center — not through tanks or terror, but through accusation. He sowed doubt, whispered slander, and weaponized free will. According to Revelation 12:7–9, there was war in heaven — a real conflict. But this wasn't just a physical expulsion; it was the beginning of a celestial legal case, a war of ideologies.

And God, in that moment, could have crushed Lucifer instantly. But He didn't. Why?

Because the combat strategy wasn't to suppress the rebellion — it was to expose it.

If God had destroyed Satan on the spot, fear would have replaced love in the hearts of His creation. Obedience would become compliance. Worship would become self-preservation. God's integrity had to be proven, not imposed.

So He allowed the rebellion to run its course — within controlled parameters, like a battlefield designed by an omniscient General.

Then came Earth — the stage of the great controversy. And Adam and Eve, like young soldiers misled by misinformation, fell into enemy hands. The plan of redemption went live.

Not created in that moment — revealed.

I think about military campaigns where a strategy had to evolve — not because the plan changed, but because the enemy made a move that revealed more of its hand. That's how God moves through prophecy. Daniel's vision of kingdoms rising and falling like metals in a statue (Daniel 2) isn't just history foretold — it's proof that God is outthinking every empire before it exists. Babylon. Persia. Greece. Rome. Divided nations. Each foretold with uncanny accuracy. Why?

Because God is not guessing. He's orchestrating.

And then Revelation gives us the battlefield: dragons and beasts, plagues and judgments, a sealed remnant, a cosmic harvest. It may sound symbolic, but it is profoundly tactical. Every piece fits — and every move made by the enemy is anticipated and countered in advance.

Satan deploys deception? God raises truth-bearers. The enemy corrupts worship? God commissions three angels with a global message (Rev 14).

A false system of religion gains ground? God prepares a remnant who "keep the commandments of God and have the faith of Jesus."

God's strategy is always a step ahead.

But here's the sobering reality — a good strategy can only save the soldier who follows it.

In war, the greatest casualties are not from bad orders — but from ignored ones.

I've seen Marines with the best equipment fall in simulations because they failed to follow their training. They panicked. They improvised. They broke from the plan.

Likewise, in the Christian life, it's not always sin that ruins people. Sometimes, it's deviation. They know the truth. They've been briefed. But they abandon formation, chase worldly comforts, or refuse to trust the Commander's orders.

- Christianity gives us a strategy. A clear one. Not to earn salvation, but to stay aligned with the divine battle plan.
- The Sabbath reminds us of the Creator in a world that worships productivity.
- The sanctuary doctrine shows us that Jesus is not idle — He's interceding.
- The health message keeps the body fit for spiritual discernment.
- The Second Coming keeps the warrior alert.
- The investigative judgment keeps the soul humble.
- This isn't legalism. It's alignment with the Commander's orders.

In military terms, we'd call it rules of engagement — not to restrict, but to ensure that the mission is accomplished and the soldier survives.

And here's where it gets personal.

Your life is a battlefield. Not metaphorically. Literally.

There are forces — both external and internal — fighting for your mind, your allegiance, your worship. Satan has studied the terrain of your weaknesses. He's laid traps, timed distractions, and disguised snares. And if you do not have a strategy, you will not survive.

You need a battle plan. Daily communion. A study regimen. Accountability. Rest. Purpose. Fellowship. Scripture intake. Worship. Fasting. Service. Mental clarity. Obedience. Stillness.

Not optional. Tactical.

Because if the enemy can't defeat you with brute force, he'll settle for disorientation. And disoriented soldiers become ineffective, isolated, and eventually... casualties.

But here's the good news: God has not left you uninformed.

The Bible is not just a devotional guide. It is a strategic combat manual. It gives you enemy patterns, divine responses, operational principles, and the final victory plan.

Jesus doesn't just call you to faith — He equips you for warfare.

And He's already won.

At the cross, Christ absorbed the full force of Satan's arsenal. Lies. Accusations. Shame. Fear. He took every bullet, every explosive, every spiritual virus the enemy could unleash. And in dying — He disarmed them.

Colossians 2:15 declares: "And having disarmed the powers and authorities, He made a public spectacle of them, triumphing over them by the cross."

In military terms? Jesus executed the ultimate flanking maneuver. The enemy thought the cross was a defeat — it was a trap. And when Jesus rose, He didn't just win — He rewrote the entire theater of operations.

Now, the war isn't about whether Christ wins. The war is about whether you'll stand under His banner or be caught in enemy territory when the final trumpet sounds.

Divine Doctrine and the Commander's Voice

In the military, doctrine is not a suggestion—it is the strategy made visible, the mind of the commander put into practical application. It's not merely about procedures or policies; it is the backbone of how battles are fought and won. We studied doctrine relentlessly at the U.S. Navy War College, not just to understand how to wage war, but to understand why certain decisions were made in the face of chaos. Doctrine keeps you aligned when the battlefield goes black. Doctrine stabilizes you when adrenaline screams at you to improvise. In combat, emotions can mislead. Panic kills. But doctrine—if you know it, believe it, and trust it—can save your life.

The same is true in the Christian life. Doctrine is not a religious burden. It is the mind of God revealed to the believer, offered not to complicate faith but to protect it. Too many Christians treat doctrine like dry theory or optional knowledge for theologians. They read the Bible selectively—skimming verses that affirm comfort and ignoring those that demand allegiance. But soldiers don't treat the Commander's words as optional. They don't reinterpret field orders based on preference or popular opinion. They follow them precisely—because they know that lives depend on it. Doctrine is divine strategy for spiritual survival. In a war where deception is the enemy's first weapon, truth must be our first defense.

I've often been asked why I embrace Adventist theology so deeply. My answer is always the same: because it is a complete, coordinated strategy for the final war. It's not a random assembly of beliefs or cultural traditions—it's a comprehensive worldview that answers the core questions of cosmic conflict, human destiny, and divine character. Each doctrine is a tactical piece of a broader campaign. Remove one, and the structure weakens. Ignore one, and the enemy finds an opening.

Take the sanctuary doctrine, for example. To most Christians, the idea of a heavenly sanctuary seems obscure, irrelevant, or ceremonial. But to those who've felt the weight of guilt, who've battled shame and lost comrades to moral collapse, the sanctuary is everything. It tells us that Christ is not a retired Savior. He is actively engaged in a phase of redemption most of the world has forgotten—His intercessory ministry. He is not idle. He is standing in our place, carrying our record, presenting His righteousness where ours has failed. Hebrews 4:15-16 reminds us that we do not have a High Priest who is detached from our pain but one who sympathizes, who invites us boldly to the throne of grace. That's not theory. That's rescue.

In my own moments of darkness—whether on deployment, in chaplaincy counseling, or in personal failure—it wasn't religious routine that kept me anchored. It was the knowledge that Christ had not given up on me. That He was interceding when I was too tired to pray. That He was presenting evidence of His mercy when the devil had plenty of evidence for my condemnation. That is doctrine—but it is also divine strategy. It silences the accuser. It strengthens the weary soldier. It reframes the war from fear to confidence.

Then there is the Sabbath. To the untrained mind, it's just a day of rest. To the shallow critic, it's legalism. But to the trained warrior, it is a countercultural act of spiritual warfare. In a world addicted to speed, consumption, and performance, to stop and rest in God is more than obedience—it's defiance. Every Sabbath, we declare that our identity is not in our productivity but in our creation. That our value isn't in what we do but in whom we belong to. In my years in uniform, Sabbath moments became holy oxygen. Whether I was alone on the ship, standing silently on the deck looking at the ocean's endless horizon, or kneeling quietly beside my bed, I remembered that the God who made time also made rest. And in that rest, I heard His voice again.

But the Sabbath is more than restoration—it's a marker. In Revelation, it becomes a line in the sand. The last great conflict isn't about morality alone—it's about worship. Whose authority do we recognize? Whose sign do we bear? The Sabbath, as the seal of the Creator, will distinguish those who follow God's law from those who follow man's traditions. It is not merely about the seventh day. It's about the strategy of allegiance. And in the final war, allegiance will determine destiny.

And what of prophecy? What of the Three Angels' Messages in Revelation 14, the hallmark of Adventist end-time proclamation? These are not esoteric warnings. They are spiritual intelligence reports. The first angel calls humanity to fear God and give Him glory—to worship the Creator as the hour of His judgment arrives. The second angel identifies Babylon—the global system of confusion, compromise, and false religion that intoxicates the world. The third angel warns against receiving the mark of the beast, the sign of loyalty to a counterfeit power. These messages are God's final call before the battle breaks into full view, before lines are drawn and loyalties sealed.

Too many believers ignore prophecy because it unsettles them. But real soldiers don't reject intel because it's uncomfortable—they lean in. Prophecy is God's GPS in the fog of war. It tells us where we are on the battlefield and where we are heading. Without it, we're fighting blind. Revelation is not a riddle—it's a briefing. A tactical forecast. A divine revelation that pulls back the curtain on the enemy's plans and God's final victory.

I've had moments in my ministry—especially in deployments—where the intensity of moral, emotional, and spiritual pressure felt unbearable. I've sat with Marines wrestling with suicide. I've counseled sailors who've lost all sense of purpose. I've stood in hallways praying over men and women whose souls were drowning in darkness. And in those moments, I realized: if I didn't know the plan, I couldn't help them trust it. My confidence in God's strategy was what gave them hope. Not vague optimism. Prophetic clarity. The knowledge that this war has an end. That every wrong will be made right. That judgment is not just accountability—it is vindication.

Doctrine is what makes you unshakable in a shaking world.

The Apostle Paul warned Timothy that in the last days, people would abandon sound doctrine, preferring myths and emotional affirmation (2 Timothy 4:3–4). Why? Because doctrine demands discipline. It demands submission. It reminds you that the Commander has already spoken—and your job is not to reinvent the strategy but to trust and follow it. In the end, that's what separates the survivors from the casualties. Not passion. Not sincerity. But alignment.

And alignment is everything in combat.

That's why I study. That's why I teach. That's why I preach what I preach. Because I'm not just trying to help people feel better. I'm trying to help them survive. To stand. To win. And doctrine—God's doctrine—is the only strategy that guarantees both victory and redemption.

The Weapons of the World

Every soldier understands that no strategy, no matter how sound, can be executed without the right equipment. Doctrine alone is not enough.

CHAPTER 14: COMBAT STRATEGIES

You must be armed, not just instructed. You can study battle maps and memorization tactics all day long, but without your rifle, your body armor, your helmet, and your training kicking in when the bullets start flying—you're vulnerable. Strategy must be coupled with strength. Planning must be accompanied by preparation. In the Christian life, the battlefield is spiritual, but the weapons are just as real—and just as essential.

Paul understood this when he wrote to the believers in Ephesus. He wasn't speaking from theory or theological detachment. He was writing as a man who had been imprisoned, beaten, hunted, and hated. A man who had endured real physical suffering for the sake of the gospel. And in Ephesians 6, he lays out the most detailed description of the believer's combat gear: the armor of God. Not as metaphor alone, but as necessity.

"Put on the full armor of God," Paul writes, "so that you can take your stand against the devil's schemes." The word "schemes" in Greek is methodia—methods, strategies, traps. The enemy is not reckless. He has a playbook. He studies your patterns. He analyzes your temperament, your habits, your wounds. That's why casual Christianity is such a dangerous thing. You may believe in Jesus, but if you are not armed—if your spirit is exposed—you are a target. And like any sniper, Satan doesn't waste bullets.

So what are the weapons we carry?

First, the belt of truth. In war, belts weren't just decorative—they held everything together. In the Christian's arsenal, truth is the stabilizer. It's what keeps you from unraveling when culture shifts and lies become fashionable. When your identity is under attack and the world is shouting contradictory messages, it is truth that secures you. Not opinion. Not comfort. But truth. And truth is found in the Word—not in feelings. I've had days when I didn't feel like I belonged to God. Days when doubt whispered louder than assurance. But it was Scripture that re-tethered me. Not my emotions. Truth doesn't depend on how I feel. It depends on who God is.

Then there's the breastplate of righteousness—covering the heart, the core of one's being. This isn't self-righteousness or religious pride. This is Christ's righteousness, gifted to us, protecting us from condemnation. It guards the heart against shame, against despair, against the fiery accusations of the enemy. And believe me, the accusations are constant. "You're not worthy." "You've failed too many times." "God is done with you." These aren't thoughts—they are darts. And if you don't have righteousness covering your chest, they will pierce you. But when you know that your worth is not earned but imputed by Christ, you walk into battle fearless.

The shoes of the gospel of peace are next. In warfare, mobility is everything. A soldier without solid footing is a liability. And the gospel gives you that footing. It grounds you in peace—not peace as absence of conflict, but peace that holds steady in the midst of chaos. The peace of knowing that your eternity is secured. That your soul is anchored. That no matter what happens in the world, you are not at war with God. You have been reconciled. That peace makes you agile in combat. You're not distracted by existential doubt. You're focused. You're steady. You're ready to move.

Then, Paul says, "take up the shield of faith, with which you can extinguish all the flaming arrows of the evil one." Faith is the most misunderstood weapon in the Christian's arsenal. It's not blind optimism. It's not naive hope. It's trust forged through battle. Faith says, "I know who God is, even when I don't see what He's doing." It blocks despair. It repels cynicism. It keeps you advancing when every circumstance tells you to retreat. And it doesn't eliminate the arrows—it extinguishes them. That means they'll still come. The thoughts. The fears. The temptations. But with faith, they lose their fire.

The helmet of salvation protects the mind. And if there's one battlefield Satan loves more than any other, it's your thoughts. The lies are relentless: "God doesn't love you." "You'll never change." "This battle is too big." "What's the point of trying?" But the helmet reminds you of who you are. Saved. Chosen. Redeemed. Bought at a price. When you wear salvation like a helmet, your thoughts begin to align with your identity. And clarity returns.

And finally, the only offensive weapon mentioned—the sword of the Spirit, which is the Word of God. This is not poetic language. This is military precision. The Word is not just for study. It is for combat. Jesus Himself used it when Satan attacked Him in the wilderness. He didn't philosophize. He didn't debate. He quoted Scripture. With every temptation, He said, "It is written." That's not superstition—that's strategy. And if the Son of God needed the Word to win a battle, so do you.

This is why I insist on daily Bible engagement—not as a chore, but as weapon maintenance. You don't take your rifle into the field without checking its chamber. You don't go on patrol with a dirty barrel. So why would you enter the day without sharpening your sword? Scripture is not just for study groups. It's for survival.

Prayer is also combat. It's not a passive ritual—it's communication with Command. It's how we receive new orders. It's how we call in reinforcements. It's how we confess our position, acknowledge our weakness, and get realigned. In war, comms are sacred. And in the Christian war, prayer is your frequency with heaven. The enemy knows this. That's why distraction is one of his greatest tools. If he can interrupt your prayer life, he can isolate you. And isolated soldiers don't last long in combat.

The Holy Spirit is the silent Operator in all of this. He guides, convicts, empowers, and teaches. He whispers warnings. He provides wisdom. He gives spiritual gifts to strengthen the body. He intercedes with groanings too deep for words. The Spirit is not an add-on. He is the active presence of God in the combat theater of your life.

I've seen too many fall not because they didn't believe—but because they weren't armed. They knew the mission but not the method. They had convictions but no weapons. Their hearts were sincere, but their hands were empty. And when the day of evil came—as Paul said it would—they were not able to stand.

This is not a drill. We are in a war. Not against flesh and blood, but against principalities, powers, ideologies, and darkness. And the only way to stand is to suit up daily. To know your weapons. To trust the Commander. To fight with precision, not panic.

There's a phrase often repeated in the field: "You don't rise to the occasion—you fall to the level of your training." And in the spiritual realm, that truth is even more urgent. When crisis hits, you won't default to inspiration. You'll default to discipline. To the Word you've stored. To the prayers you've prayed. To the truth you've trained in.

So train well. Fight smart. Stay armed.

Because the war is real.

And you were born to win it.

The Victory That Rewrites the War

Victory is a strange thing in war. It's not always the loudest moment. Sometimes it's a whisper—a decision made in a command center, a signal received, a flag raised quietly over smoking rubble. It often comes with grief mingled with relief, because victory in war is always stained with blood. And yet, in every conflict, there comes a point when the outcome shifts. When the tide turns. When one side seizes control and everything changes after that.

In spiritual warfare, that moment happened two thousand years ago on a hill called Golgotha.

It looked nothing like victory. A naked man, beaten and mocked, nailed to a wooden beam between two criminals. No military band. No parade. Just blood, cries, and jeers. If any of us had stood there as observers, we would have assumed the enemy had won. That evil had triumphed. That God had failed.

But heaven saw differently. Because the Cross wasn't a mistake in the strategy—it was the climactic maneuver. The place where the ultimate flank was executed. Where Satan poured every weapon he had onto Jesus —betrayal, injustice, torture, abandonment—and still failed to stop Him. In fact, he walked right into a divine ambush.

Colossians 2:15 declares: "Having disarmed the powers and authorities, He made a public spectacle of them, triumphing over them by the cross."

Triumph—not at the resurrection, but at the Cross. That was the moment the war turned. Satan's credibility collapsed. His accusations fell flat. The curse was broken. And though battles would still rage, the outcome had been sealed.

From a military standpoint, it was the perfect counterintelligence operation. God allowed the enemy to exhaust every resource, convinced he was winning, only to reveal that the Cross was never defeat—it was a trap. And Christ, by dying, detonated the scheme. The grave was shattered from within. And when Jesus walked out of the tomb, He didn't just rise—He redefined war itself.

That's the God we serve. A Commander who bleeds. A King who rescues His soldiers not by issuing commands from a throne, but by stepping into the front line and absorbing the fire Himself. And now, everything we do—from preaching to parenting to resisting temptation—is not to earn victory, but to stand inside of one already won.

But if Christ's death was the turning point, then His Second Coming is the final assault—the decisive strike that ends the conflict forever.

Christian eschatology makes this clear. We don't believe the world will gradually evolve into utopia. We don't believe humanity will fix itself. We believe in a returning King. One who will interrupt history. One who will appear not as a carpenter or servant, but as a warrior on a white horse, eyes like fire, sword from His mouth, crowned with many crowns (Revelation 19).

This is not myth. It's the final move in a strategy planned before the foundation of the world. The Second Coming isn't just hope—it's culmination. It is the visible demonstration of Christ's rightful authority, the rescue of His faithful, and the destruction of every system that opposed truth.

When I studied at the Navy War College, one of the most sobering lessons was the doctrine of total war—a state in which the full resources of a nation are engaged in conflict, and the separation between battlefield and civilian space disappears. In many ways, that's the world we're in now. Sin has made everything a battlefield—your relationships, your mind, your choices, even your rest.

There's no such thing as neutrality anymore. You're either aligned with the Kingdom or being slowly absorbed by the system of Babylon. You're either marching toward Zion or lounging in Sodom. And the only way to know where you are is by your allegiance, your weapons, and your orders.

That's why prophecy matters. Not for fear, but for formation.

Daniel 7 shows us thrones set in place, a judgment scene, and the Son of Man receiving the kingdom. Revelation reveals beasts, false prophets, plagues, and a sealed remnant. These aren't bedtime stories—they are operational briefings. God isn't trying to scare us into obedience—He's training us to discern, to endure, and to trust.

Some scoff at the doctrine of the investigative judgment—"Why would God need a judgment if He knows everything?" they ask. But they miss the point. Judgment isn't for God's information. It's for the universe's vindication. It's the final piece of His strategy: to prove that every saved soul was not just forgiven—but transformed. That grace didn't just pardon—it recreated. That Satan's accusation—that God's law is unfair and His grace is unjust—has no ground to stand on.

Every detail fits the plan.

And that plan ends with justice. Not revenge. Not genocide. But justice. Holy, fair, irreversible justice. And for those who trust the Lamb and walk in His ways, the justice of God is not a threat—it's a rescue.

That's why I fight.

Not because I fear losing, but because I want to be found standing when the sky cracks open. I want to be among those who say, "This is our God; we have waited for Him!" (Isaiah 25:9). I want my life to be a statement of allegiance—not just in word, but in formation, in sacrifice, in trust. And I want others to find their place in this army—not by force, but by love. Not through shame, but through revelation.

The redemption plan is the greatest combat strategy ever written. A campaign of cosmic proportion. A war of truth and deception. A battle between love and pride. And we, right now, stand in the final phases.

I often wonder what it would have been like to be at Midway. To be on the deck when the tide turned. To realize the enemy was retreating. To hear the crackling of the radio as the orders shifted from defense to pursuit.

But I don't have to wonder. Because I'm living it.

We're in the final stretch of a war already won but not yet finished.

And we have a role to play. A message to bear. A people to become.

So soldier on. Suit up. Stay sharp. Know the plan. Follow the Commander. And never forget that you are not fighting for victory. You're fighting from it.

CHAPTER 15

The Merciless Enemy Attacks

THERE ARE PLACES ON THIS EARTH where the stench of death outlasts the smoke of war. Where the screams of mothers and the moans of the dying linger in the soil long after the gunfire ends. Places where the names of cities have become synonymous with suffering —Auschwitz. Kigali. Fallujah. Srebrenica. My Lai. Bucha. And behind each of those names are faces—men, women, and children—whose lives were mutilated by forces so evil, so deliberate, so inhumane, that the only appropriate word is merciless.

Where the Darkness Lives

Evil does not act out of impulse. It strategizes. It waits. It studies your weakness and then unleashes destruction with surgical precision. That is the nature of our physical enemy in war—and it is the exact nature of our spiritual enemy in the great controversy. The devil doesn't just tempt the world into sin. He incites genocide. He whispers commands into the hearts of tyrants. He distorts ideology into hate. He hardens hearts until the murder of thousands becomes not only acceptable, but rationalized. This isn't hyperbole. It's history.

In 1994, nearly a million Tutsis were slaughtered in Rwanda in just 100 days. Some were hacked with machetes. Others burned alive. Churches became mass graves. Friends killed neighbors. Teachers betrayed students. It wasn't spontaneous. It was orchestrated. And it was ignored by most of the world until it was too late.

In World War II, six million Jews were murdered in the Holocaust. Not by wild mobs, but by scientists, engineers, doctors, and civil servants—educated men who turned human extermination into policy. Who weighed children to determine their "usefulness" and built gas chambers with calculated efficiency. It was not chaos. It was the chilling result of what happens when evil is left unopposed.

You cannot read these accounts and believe in the inherent goodness of mankind. You cannot study the genocides of Armenia, Cambodia, Darfur, or the horrors of ISIS and Boko Haram, and still say evil is a myth or the military is unnecessary. War is not always the result of aggression. Sometimes, it's the only response to evil.

That's why I chose to wear the uniform.
Not to conquer. Not to dominate. But to stand between the predator and the prey. Because when I looked at what evil had done to the world, I realized something profound: it's not enough to be sad about injustice—you have to be willing to stand against it. And in some moments, standing requires more than protest signs and prayer meetings. It requires armor. It requires action. It requires Marines.

There is a reason Paul describes our spiritual life in terms of warfare. Because evil is not passive. It doesn't wait for permission to act. It doesn't play by rules. It doesn't respect treaties. Satan doesn't take holidays. He doesn't spare the innocent. He doesn't offer mercy. He invades, he corrupts, and he seeks to devour. As 1 Peter 5:8 says, "Be sober-minded; be watchful. Your adversary the devil prowls around like a roaring lion, seeking someone to devour." This isn't figurative language. It's an intelligence briefing. And the sooner we wake up to the reality of that

CHAPTER 15: THE MERCILESS ENEMY ATTACKS

enemy, the sooner we can stop being victims and start becoming warriors.

I've served in conflict zones. I've walked into villages where the air was still heavy from mortar fire. I've looked into the eyes of children whose parents were executed the night before. I've stood over the bodies of soldiers who didn't make it. Evil is not a philosophy to me. It's a smell. It's a sound. It's the absence of light in places where hope should live.

But here's the deeper truth: the darkness in this world is a mirror of the greater war that rages unseen. Behind every massacre is a spirit. Behind every atrocity is an agenda. Behind every tyrant is a whisper. And behind it all is a single objective: to mock the image of God in man. To destroy what He created. To undo what Christ redeemed.

Satan doesn't just want your sin—he wants your surrender. He wants you to believe that evil is inevitable. That goodness is powerless. That war is always unjustified. Because if he can convince you of that, then he can keep you from standing up when standing matters most.

This is why evil flourishes—not because the wicked are strong, but because the righteous are silent.

But silence is not an option for the believer. Not in prayer. Not in war. Not in defense of the innocent. We don't serve a passive God. We serve the God who confronted Pharaoh, who sent fire on Mount Carmel, who overturned tables in the temple, who will one day return with a sword in His mouth and justice in His hands. Evil fears that day. And it should.

The military exists not as a tool of power, but as a barrier to chaos. When placed in the hands of the righteous, it becomes a restraint on darkness—a visible reminder that evil will not go unanswered. And when that military is filled with men and women who know God, who serve under the ultimate Commander, it becomes more than an army—it becomes a force for divine justice.

But we must be honest: the presence of good soldiers doesn't erase evil. Sometimes, the Marines show up after the massacre. Sometimes, we bury more than we rescue. Sometimes, the darkness feels too thick. That's when the spiritual battle intensifies. That's when faith becomes the only thing left to hold onto.

And it is there—in the ruins, in the grief, in the chaos—where I have found God to be the most present.

Not because He willed the violence. But because He never abandoned the wounded.

Because while the devil orchestrates atrocity, God rebuilds lives in the aftermath. He brings survivors together. He heals minds destroyed by trauma. He speaks into the silence left by evil's scream. And He raises up warriors—men and women who say, "Not again. Not here. Not on my watch."

This is the heart of the Christian Marine. One who is not blind to the world's darkness—but refuses to let it go unchallenged.

Why Warriors Must Exist

There are moments in history where hesitation becomes complicity. Moments when the cost of inaction is paid in civilian blood and silence becomes the devil's greatest accomplice. Evil does not need everyone to participate—it only needs the righteous to stand down. And in a world where genocide, terrorism, human trafficking, and political oppression still disfigure the human soul, the existence of warriors is not only justified—it is required. It is biblical.

The concept of war makes many Christians uncomfortable, and rightly so. War is never clean. It tears at the fabric of humanity, dismantles families, ruins economies, and breeds generational trauma. But sometimes, what is uncomfortable is still necessary. To say that evil exists is not the same as saying war is good. But to deny the need for warriors is to deny the reality that some evils will never stop unless someone makes them stop.

Scripture does not glorify violence, but it is unflinchingly honest about its place in a fallen world. In Ecclesiastes 3:8, Solomon wrote that there is "a time for war, and a time for peace." Not every time is war, but not every time is peace either. The same God who said "blessed are the peacemakers" is the One who told Joshua to conquer Jericho, who called Gideon to lead an army, and who empowered David—a warrior-king who spilled blood in defense of God's people.

Romans 13:4 goes even further. Paul, writing under the inspiration of the Holy Spirit, says of the governing authority: "For he is God's servant for your good. But if you do wrong, be afraid, for he does not bear the sword in vain." This is not an endorsement of tyranny. It's an acknowledgment that force, when under righteous command, is a tool of divine justice.

This doesn't mean every war is right. History proves otherwise. Some wars are born of pride, greed, nationalism, or vengeance. Those wars stain the hands of both the attacker and the silent witness. But the existence of unjust wars does not negate the legitimacy of just ones. The fact that some have abused power does not mean that power itself is evil. It means that power, like fire, must be wielded by those who fear its consequences and respect its purpose.

This is why the training of a Marine is more than just a matter of physical readiness. It's moral formation. It's learning how to restrain the power you've been entrusted with. It's understanding when to act and when to hold back, when to fire and when to fall back, when to lead and when to protect. The line between warrior and war criminal is not the weapon in their hands—it's the compass in their soul.

And here's where faith becomes the ultimate compass.

A Marine without a moral anchor may be brave, but bravery without righteousness is just reckless aggression. A Christian Marine, on the other hand, understands that justice is not vengeance, and power must serve purpose. He does not kill for pride. He does not destroy for gain. He does not relish in the chaos of conflict. Instead, he steps into the fight because someone must. And he fights with the weight of eternity on his shoulders.

I've seen Marines stand between a woman and her abuser, risking their life not because it was in their orders, but because it was right. I've seen them carry wounded children through warzones because the mission wasn't just to take a hill, but to protect the innocent. I've stood beside men who cried after firefights, not from fear, but from grief—because taking life, even when justified, should never feel casual.

This is what separates warriors from killers. A warrior knows what he's fighting for.

We must remember: evil is not defeated by passivity. When ISIS swept through villages in Iraq and Syria, enslaving women, executing Christians, and indoctrinating children to become suicide bombers, the world watched in horror. It took force—righteous, strategic, and unrelenting force—to stop them. And yes, the gospel must go to all nations, including our enemies. But until they are willing to listen, someone must shield the victims of their hatred.

Pacifism in the face of genocide is not nobility—it's negligence. Neutrality in the face of rape camps and concentration camps is not virtue—it's cowardice dressed in moral confusion. Jesus Himself did not remain passive. When injustice filled the temple, He didn't host a prayer circle—He flipped tables. When Satan tempted Him in the wilderness, He didn't negotiate—He declared truth with force. And when He returns, He won't come as a baby in a manger. He'll return as a warrior-King, with justice on His lips and fire in His eyes.

Revelation 19 paints the picture: "Then I saw heaven opened, and behold, a white horse! The one sitting on it is called Faithful and True, and in righteousness He judges and makes war." That's not poetic flair. That's prophecy. That's Jesus, leading the final charge to destroy evil forever. If the Son of God makes war in righteousness, how can we pretend war is never needed?

To deny the need for warriors is to pretend this world is already Eden. It's not.

We live in a world where traffickers still make money from the sale of human bodies. Where entire governments profit from oppression. Where children are given weapons instead of education. In such a world, warriors must exist—not because war is good, but because evil is real.

But make no mistake: not every soldier is a warrior. And not every warrior is submitted to God. The world is filled with men who love war for the wrong reasons. But that only means the need for righteous warriors is even greater.

Men and women who will fight with clarity, not hatred. Who will defend the innocent, not feed their ego. Who will bear arms with trembling hands and grounded hearts.

Who will know when to fight, when to kneel, and when to weep.
Because they serve not just a nation, but a King.

Satan, the Invisible Tyrant

There is a reason the Bible so often compares Satan to rulers and tyrants. He doesn't come as a shadowy figure in a haunted forest. He comes like Pharaoh, refusing to let the oppressed go free. He comes like Nebuchadnezzar, demanding worship. He comes like Herod, slaughtering innocents to protect his seat. Like Caesar, like Hitler, like any number of kings, presidents, and generals who abused their power—Satan governs through fear, deception, and control.

But he doesn't always use bullets or bayonets. Sometimes, the most devastating attacks are waged in the shadows of the mind.

As Marines, we are trained to identify a threat. To observe the terrain. To know the enemy's tactics, weaknesses, and strongholds. That's how wars are won—through awareness, preparation, and precision. But many Christians step into spiritual battle every day completely unaware they're even in a war. They don't recognize the sniper fire of shame. They miss the mines of compromise. They don't hear the enemy's propaganda whispered into their thoughts: You're not enough. You've failed too many times. You'll never change.

Satan doesn't need to make you a murderer to win. He just needs to make you neutral. Passive. Spiritually unarmed.

But Marines don't get to be neutral. And neither do believers. Not when souls hang in the balance. Not when the enemy is assaulting our families, distorting truth, redefining morality, and poisoning the minds of entire generations. We don't get to pretend this is peacetime when casualties are all around us—marriages imploding, young people abandoning faith, addictions taking root, and hope growing dim.

We are at war.

And our enemy does not play fair.

In the military, we study the psychological warfare of past conflicts. The brainwashing of POWs in Korea. The propaganda machines of Nazi

Germany. The fear-based control of totalitarian regimes. Satan has perfected all of these techniques on a spiritual level. He doesn't just oppose truth—he twists it. He doesn't just tempt you with sin—he makes it look righteous. He doesn't just attack the church—he joins it, waters down the message, and cloaks apathy in religious language.

Jesus described Satan not simply as evil, but as "a liar and the father of lies" (John 8:44). Deception is his native tongue. And unless we're fluent in truth, we'll be manipulated without even realizing it.

This is why the Christian Marine must go beyond physical readiness. You can have a chiseled body, perfect marksmanship, and combat instincts honed in the field—and still lose if you are spiritually disarmed. This war is not just about IEDs and insurgents. It's about ideology. Identity. Eternity. If you don't know who you are in Christ, Satan will hand you a false identity and convince you it's real.

I've counseled Marines who looked invincible on the outside—but inside were drowning in shame, guilt, pornography addiction, suicidal thoughts, bitterness, or pride. Their war wasn't just on deployment. It was in their dorm rooms. In their silence. In their unresolved trauma. And without the armor of God, they were vulnerable—even as warriors.

Paul's letter to the Ephesians wasn't written to civilians. It was written to the front lines of the faith. That's why he said: "Put on the full armor of God, so that you can take your stand against the devil's schemes…" (Ephesians 6:11).

Notice the word schemes. Satan is not improvising. He's orchestrating. He's studying your patterns. He's planning your ambush. That moment of temptation isn't random. It's a sniper shot planned weeks in advance. That offense you can't let go of—it's a trap set to isolate you. That secret sin you won't confess—it's a spiritual minefield waiting to detonate.

This is why faith without awareness is dangerous. If you don't realize you're being hunted, you won't take cover. You won't check your perimeter. You won't stay alert. And eventually, you'll become collateral damage—another casualty in a war you weren't paying attention to.

But here's the good news: our Commander has already exposed the enemy's tactics, and He has not left us defenseless.

CHAPTER 15: THE MERCILESS ENEMY ATTACKS

We have weapons that the world cannot see—prayer that moves angels, truth that demolishes strongholds, righteousness that covers shame, salvation that secures the soul, Scripture that cuts deeper than bullets, and the Spirit who never retreats.

Satan may be ruthless, but he is not creative. He keeps using the same lies, the same tools, the same traps. And when we train ourselves in truth —when we stay in the Word, stay in community, stay on our knees—we don't just survive. We advance. We become a threat.

This is what the enemy fears most—not loud Christians, but rooted ones. Not emotional Marines, but anchored ones. Not religious talk, but spiritual readiness. When a believer knows who they are, whose they are, and why they've been placed on this battlefield, there is nothing more dangerous.

Because the Christian Marine isn't just fighting for survival. He's fighting for souls. He's fighting for the oppressed. He's fighting for integrity in a corrupt world. And he's not doing it alone.

We fight under the banner of the One who already won the war.

Satan may rage. He may attack. But every one of his weapons has already been defanged by the Cross. His authority is counterfeit. His accusations are hollow. And his future is sealed.

That doesn't mean the war isn't real. But it does mean we fight from victory, not for it.

And when the tyrant rises, we don't run.

We remember who we serve.

The Christian Marine

There's a quiet moment just before a mission begins—after the final gear check, after the last briefing, when the adrenaline is still waiting for permission to surge. It's in that silence, in that sacred stillness, that you get to ask yourself the most important question a Marine ever will: Why am I doing this?

Some fight for the man beside them. Some fight for their country. Some fight because they love the war. But I fight because I believe I was born for this time. Not by accident. Not by politics. But by calling. I believe I was placed here—uniform, rifle, boots, conviction—not just to defeat enemies, but to represent Christ in a world that doesn't know what righteousness in warfare looks like.

This is what sets a Christian Marine apart. Not just training. Not just toughness. But transcendence—the awareness that we are not just part of the Corps… we are part of a Kingdom.

A Marine without faith can fight bravely, no doubt. He may serve honorably. He may make sacrifices. But the Christian Marine sees the battle through a different lens. Every conflict is a spiritual crossroads. Every life—civilian, enemy, or ally—is eternal. Every choice made in the fog of war ripples beyond this life and into the next. We carry more than ammo. We carry the presence of God.

This doesn't make us perfect. In fact, it makes us more aware of our imperfections. The weight of moral clarity demands humility. And the closer you walk with God in combat, the more conscious you become of the fragility of human life, of the cost of every bullet fired, of the deep ache that justice, even when rightly carried out, is never something to celebrate flippantly.

But this spiritual awareness is also what gives the Christian Marine his strength. When you've settled the question of eternity, you can face danger differently. When your identity is rooted in Christ, insults don't break you, and praise doesn't define you. When you know that heaven is real and Christ is near, courage becomes second nature, not just because you've trained for war—but because you've trained for faith.

I've seen men with superior firepower fall apart under pressure, and I've seen Christian Marines carry others on their shoulders through gunfire, praying as they moved, bleeding but believing. Not because they were more fearless, but because they were more anchored.

You see, belief doesn't make you immune to fear. It makes you faithful in the face of it.

It teaches you to choose integrity over impulse. To resist the temptation to dehumanize your enemy. To serve with resolve, but never with hatred. To embrace the tension of being both warrior and

CHAPTER 15: THE MERCILESS ENEMY ATTACKS

worshipper, fighter and follower, disciplined and dependent.

There is a depth of moral burden in combat that no atheist philosophy can explain. When you're standing on a battlefield, facing someone who would take your life without hesitation, and yet you still feel the weight of his humanity, the tragedy of his lostness, the sacred reality that he, too, was made in the image of God—that's a moment only faith can interpret.

And that's why a Christian Marine, though trained like every other, is not like every other.

He sees further. He aims with a trembling conscience. He fights with purpose, not just pressure. He knows that every mission is a theater of spiritual consequence.

And he is held accountable not just by his CO—but by the Creator.

The Code of Conduct is etched into every Marine's mind. But for the believer, there's another code—written not in ink, but in blood. The blood of the Lamb. The sacrifice of Christ. The calling to live not just with honor, courage, and commitment—but with holiness, justice, and love.

This doesn't make the Christian Marine superior. It makes him more responsible. Because with revelation comes responsibility. With faith comes burden. With divine backing comes divine accountability.

And sometimes, that burden isn't in pulling the trigger—it's in holding your fire. Sometimes it's in praying for the man who tried to kill you. Sometimes it's in comforting a civilian child who now fears your uniform. Sometimes it's in holding your brothers accountable when they're about to cross a moral line.

This is the unseen warfare that many never talk about. The internal combat. The moral crossroads. The secret prayers whispered beneath a helmet. The silent confessions after a mission. The constant tension between warrior instinct and godly restraint.

The Christian Marine doesn't just survive this tension. He thrives in it —because he understands the mission is bigger than territory. Bigger than medals. Bigger than glory. It's eternal.

He fights not just for freedom, but for truth. Not just for America, but for the Kingdom. Not just for victory, but for the testimony his life will leave long after the war is over.

And when the war is done—when the deployment ends, when the uniform is folded, when the battlefield fades—he doesn't hang up his faith with his gear. Because for the Christian Marine, the fight doesn't end with discharge. It ends with the return of the King. Until then, he fights. He prays. He serves. Not perfectly, but faithfully. Not loudly, but powerfully. Not for himself, but for the glory of the One who gave His life to rescue enemies and turn them into sons.

Fighting With Christ Beside You

There is a moment in every battle—spiritual or physical—when the ground seems to vanish beneath your feet. When the plans unravel, the radio goes silent, and all that remains is instinct. It's in these crucible moments that a Marine discovers what truly anchors him. For some, it's rage. For others, revenge. But for the believer, it is presence—the unshakable awareness that Christ is beside him, even in the gunfire.

This is not poetic optimism. This is reality. When Jesus said, "Lo, I am with you always, even unto the end of the world" (Matthew 28:20), He didn't mean only in church pews or quiet prayer rooms. He meant in trenches. On patrol. In the kill zone. In the hospital bay. In the suicide watch tent. He meant in the moment your squadmate bleeds out beside you. He meant in the silence after your last bullet is fired. He meant always.

And this is what changes everything for the Christian Marine. Knowing that the Savior who conquered death walks beside you into the valley of its shadow. Not as a distant deity, but as a fellow warrior, scarred by the very war He came to win—against sin, against death, against the devil himself.

In Revelation 19, Christ is described not as a gentle lamb but as a warrior on a white horse, riding with justice and fire. His robe is dipped in blood—not ours, but His own. The King who leads us is not unfamiliar with war. He fought the greatest one at Calvary, where darkness gathered all its strength for one final assault and failed spectacularly.

The Cross was not the end of a martyr's life. It was the turning point

CHAPTER 15: THE MERCILESS ENEMY ATTACKS

in the greatest campaign in the universe. And the empty tomb is our permanent battle flag—proof that evil can be resisted, and righteousness will triumph.

This is why we fight. Not because we love conflict, but because we love what the enemy is trying to destroy—innocence, dignity, family, truth, freedom, faith. Satan's hatred isn't generic. It's targeted. He wants to burn what God builds, enslave what God sets free, defile what God calls holy. That's why warfare, both physical and spiritual, is not about violence—it's about preservation.

We are not conquerors; we are guardians.

The Apostle Paul reminded young Timothy, "Suffer hardship with me, as a good soldier of Christ Jesus" (2 Timothy 2:3). This wasn't decorative language. It was the soul of discipleship. Because to follow Christ is to enlist in a war where the casualties are eternal. To be Christian is to be at odds with every force that opposes God's redemptive plan. We are not civilians in this world. We are embedded operatives—rescuers, intercessors, ambassadors in hostile territory.

And here's what I've learned from the battlefield: the most dangerous enemy is not the one who charges at you—but the one who blends in. Satan doesn't always show up in blood and fire. Sometimes he comes in comfort. In compromise. In entertainment. In distraction. He numbs your discernment before he launches his assault. And when your sword is dull, and your armor is off, he strikes.

That's why Christian warriors cannot afford to be casual. We must live ready, not just in body, but in spirit. Not just on base, but in the barracks. Not just on Sundays, but in every moment.

We must study the Word like it's a combat manual.

We must train our minds like we train our muscles—sharpened, focused, alert.

We must build spiritual endurance through prayer, fasting, community, confession.

Because the day will come—if it hasn't already—when you'll be asked to make a decision that could cost you your reputation, your rank, your comfort, or your life.

And in that moment, your soul will lean on whatever you've trained it to trust. If you've trained in self-preservation, you'll run. If you've trained in pride, you'll crumble under failure. If you've trained in compromise, you'll justify retreat. But if you've trained in truth, if you've learned to hear the Commander's voice, if you've walked with Christ in the quiet moments, then in the chaos, you'll stand.

You won't need to shout to be heard. You won't need to threaten to be respected. You'll stand because you've already died—died to sin, to ego, to fear—and now you live not by instinct, but by faith.

A Christian Marine is not superhuman. He bleeds like any other. He weeps like any other. He feels the weight of grief, the sting of injustice, the fatigue of constant tension. But what makes him formidable is that he knows how to lay that burden down at the feet of the One who understands it best.

He's not afraid of death, because he knows it's not the end. He's not undone by evil, because he knows it's already lost. And he's not shaken by failure, because he knows grace is stronger than guilt.

The Christian Marine fights, not for medals, but for the mission. And the mission is simple: to stand between the enemy and the ones who cannot yet stand for themselves. To be light in the dark. To be a protector in a culture of predators. To be a voice of truth when lies echo like thunder.

We may not win every battle. But we fight every one as those who know how it ends. We don't fight for a throne. We fight because the throne is already occupied.

We fight beside Christ. With Christ. And because of Christ.

And we fight until the day the final trumpet sounds, and war is no more.

Amen.

BIBLIOGRAPHY

Aland, Kurt, Matthew Black, Carlo M. Martini, Bruce M. Metzger, and Allen Wikgren, eds. *The Greek New Testament, 4th Revised Edition*. 4 Revised ed. Stuttgart, Germany: American Bible Society, 2000.

Association of Clinical Pastoral Education. "Accreditation." Accessed November 30, 2020. https://acpe.edu/programs/accreditation

Barna Group. *Six Reasons Young Christians Leave Church*. Ventura, CA: Barna Research Group, 2011. https://www.barna.com/research/six-reasons-young-christians-leave-church/.

Barna Group. *The Connected Generation*. Ventura, CA: Barna Research Group, 2020. https://www.barna.com/the-connected-generation/.

Bereit, Rick. In His Service: *A Guide to Christian Living in the Military*. Colorado Springs: Dawson Media, 2002.

Budde, Michael L. *The Borders of Baptism: Identities, Allegiances, and the Church*. Eugene, OR: Wipf & Stock Pub, 2011.

Centers for Disease Control and Prevention. *Youth Risk Behavior Survey: 2023 Summary & Trends Report*. Atlanta, GA: U.S. Department of Health and Human Services, 2023. https://www.cdc.gov/healthyyouth/data/yrbs/.

Kinnaman, David, and Mark Matlock. *Faith for Exiles: 5 Ways for a New Generation to Follow Jesus in Digital Babylon*. Grand Rapids, MI: Baker Books, 2019.

Burrill, Russell. *Reaping the Harvest: A Step-by-step Guide to Public Evangelism*. Fallbrook, California: Hart Books, 2007.

Campbell, I. D. *Matthew's Gospel (opening Up)*. Leominster: DayOne Publications, 2008.

Currier, J. M., Holland, J. M., Rojas-Flores, L., Herrera, S., & Foy, D. (2015). Morally injurious experiences and meaning in Salvadorian teachers exposed to violence. *Psychological Trauma: Theory, Research, Practice, and Policy*, 7, 24–33. http://doi.org/10.1037/a0034092

Dudley, Roger L. and Edwin I. Hernandez. *Citizens of Two Worlds*. Berrien Springs: Andrews University Press, 1992.

Eckerlin, D. M. , Kovalesky, A. & Jakupcak, M. (2016). CE. AJN, *American Journal of Nursing*, 116 (9), 34-43. doi: 10.1097/01.NAJ.0000494690.55746.d9.

Goldingay, J. (2009). *Old Testament Theology, Volume 3: Israel's Life*. Downers Grove, IL: InterVarsity Press, 234.

"Greek Verbs Quick Reference," last modified December 01, 2011, accessed June 22, 2013, http://www.preceptaustin.org/new_page_40.htm.

Herndon, B. (1967). *The Unlikeliest Hero* (Mountain View, CA: Pacific Press Publishing, 78.

Hipes, C., Lucas, J. W., & Kleykamp, M. (2015). Status- and stigma-related consequences of military service and PTSD: Evidence from a laboratory experiment. *Armed Forces and Society*, 41(3), 477-495.

Hoehner, H. W. (1985). Ephesians. In J. F. Walvoord & R. B. Zuck (Eds.), *The Bible Knowledge Commentary: An Exposition of the Scriptures* (J. F. Walvoord & R. B. Zuck, Ed.) (Eph 2:19). Wheaton, IL: Victor Books.

Kaiser Jr., W. C. (1983). *Toward Old Testament Ethics*. Grand Rapids, MI: Zondervan, 75.

Kittel, G., Friedrich, G., & Bromiley, G. W. (1985). *Theological Dictionary of the New Testament* (201). Grand Rapids, MI: W.B. Eerdmans.

Kopacz, M. S., Adams, M. S., Searle, R., Koeing, H. G., & Bryan, C. J. (2019). A preliminary study examining the prevalence and perceived intensity of morally injurious events in a Veterans Affairs chaplaincy spiritual injury support group. *Journal of Healthcare Chaplaincy*, 25(2), 76–88. https://doi.org/10.1080/08854726.2018.1538655

Kraybill, D. B. (2003). *The Upside-Down Kingdom*. Scottdale, PA: Herald Press, 56.

Merriam-Webster, Inc. *Merriam-Webster's Collegiate Dictionary*. Eleventh ed. Springfield, MA: Merriam-Webster, Inc., 2003. 983.

Michael Hoefer, Nancy Rytina, and Bryan Baker. "Estimates of the Unauthorized Immigrant Population Residing in the United States: January 2011." http://www.dhs.gov/xlibrary/assets/statistics/publications/ois_ill_pe_2011.pdf. June 23, 2013. Accessed June 23, 2013.

Mole, Robert L., and Dale M. Mole. *For God and Country*. Brushton: Teach Services, Inc., 1998.

Mota, Natalie, Jordana L. Sommer, Shay-Lee Bolton, Murray W. Enns, Renée El-Gabalawy, Jitender Sareen, Mary Beth MacLean, et al. 2022. "Prevalence and Correlates of Military Sexual Trauma in Service Members and Veterans: Results from the 2018 Canadian Armed Forces Members and Veterans Mental Health Follow-up Survey." *The Canadian Journal of Psychiatry*, September (September), 070674372211252. https://doi.org/10.1177/07067437221125292.

Myers, Allen C. *The Eerdmans Bible Dictionary.* Grand Rapids, MI: Eerdmans, 1987.

Niebuhr, R. (1932). *Moral Man and Immoral Society: A Study in Ethics and Politics.* New York: Charles Scribner's Sons, 189.

Pasquale, Michael, and Nathan L.K. Bierma. *Every Tribe and Tongue: a Biblical Vision for Language in Society.* Eugene, OR: Wipf & Stock Pub, 2011.

Phillips, K., & Tsatalbasidis, K. (2007). *I Pledge Allegiance: The Role of Seventh-day Adventists in the Military.* USA: Keith Phillips.

Rodas, Daniel C. *Christians at the Border: Immigration, the Church, and the Bible.* Grand Rapids, Mich.: Baker Academic, 2008.

Strong, J. (2009). *Vol. 1: A Concise Dictionary of the Words in the Greek Testament and The Hebrew Bible* (20). Bellingham, WA: Logos Bible Software.

Utley, R. J. (1997). *Vol. Volume 8: Paul Bound, the Gospel Unbound: Letters from Prison (Colossians, Ephesians and Philemon, then later, Philippians).* Study Guide Commentary Series (89). Marshall, TX: Bible Lessons International.

Volf, M. (1996). *Exclusion and Embrace: A Theological Exploration of Identity, Otherness, and Reconciliation.* Nashville, TN: Abingdon Press, 117.

Wilcox, Francis McLellan. *Seventh-day Adventists in Time of War.* Takoma Park: Review and Herald Publishing Association, 1936.

Wolstenholm, David. *Combat Ready.* Bloomington: WestBow Press, 2012.

www.ingramcontent.com/pod-product-compliance
Lightning Source LLC
Chambersburg PA
CBHW060509100426
42743CB00009B/1269